Nuclear Doping of Semiconductor Materials

LAP LAMBERT Academic Publishing

Imprint

Cover image: www.ingimage.com

Publisher:
LAP LAMBERT Academic Publishing
is a trademark of
International Book Market Service Ltd., member of OmniScriptum Publishing Group
17 Meldrum Street, Beau Bassin 71504, Mauritius

Printed at: see last page
ISBN: 978-620-0-48643-1

Vjacheslav Kharchenko

Nuclear Doping of Semiconductor Materials

INSTITUTION OF RUSSIAN ACADEMY OF SCIENCES
DORODNICYN COMPUTING CENTRE
OF RAS

NUCLEAR DOPING OF SEMICONDUCTOR MATERIALS

Second Edition, revised

Executive Editor, Doctor of Engineering Sciences,
V.A. Kharchenko

Moscow 2020

Nuclear Doping of Semiconductor Materials.

The monograph discusses the physical foundations of a new method of doping semiconductors, the basic element of which is nuclear reactions that occur in the volume of a semiconductor under the influence of fast charged particles, neutrons, high-energy gamma radiation, as well as the inevitably occurring side effects - the formation of radiation defects, the kinetics of their accumulation and annealing. Detailed data are presented on the technology of uniform irradiation of bulk ingots with neutrons depending on the specific design of the nuclear reactor, the medium and modes of annealing of radiation defects, the requirements for the starting material, and the electrophysical properties of doped silicon crystals are considered. The sources of radioactive contamination of the ingots during the irradiation process and the technological methods of their deactivation to a safe level are analyzed.

The monograph is designed for scientists and production personnel interested in the problems of solid state radiation physics and radiation materials science of semiconductors and devices based on them, as well as graduate students and students of relevant specialties.

Keywords: semiconductors, doping, nuclear reactions, silicon single crystals, neutron irradiation, radiation defects, annealing, electrophysical properties.

Author: **V.A. Kharchenko,**

L.S. Smirnov	**S.P. Solov'ev**	**V.F. Stas'**

Vyacheslav Aleksandrovich Kharchenko, engineer - physicist, doctor of technical sciences. He completed a series of research works on the development of a new technology for doping semiconductors using nuclear reactions and the creation of a production line based on the WWR-t nuclear reactor (Obninsk, Kaluga Region). Subsequently, he was engaged in the introduction of new equipment and technologies at the enterprises of the semiconductor industry. Currently, a leading researcher at the Institution of russian academy of sciences dorodnicyn computing centre of RAS. Scientific research is related to the modeling of semiconductor elements of basic heterostructures used in various electronic devices.

Smirnov Leonid Stepanovich, Doctor Phys.-Math. Sci., Rzhanov Institute of Semiconductor Physics Siberian Branch of Russian Academy of Sciences.

Soloviev Sergey Petrovich, Doctor of Phys.-Math. Sciences, Branch of the Physico-Chemical Institute named after L.Ya. Karpova, Obninsk.

Stas Vladimir Fedoseevich, candidate of physical and mathematical sciences, Rzhanov Institute of Semiconductor Physics Siberian Branch of Russian Academy of Sciences.

Contents

Preface to the second edition ... 6

Preface ... 9

Chapter 1. Doping of semiconductors ... 11

1.1. Melt doping ... 11

1.2. Impurity diffusion ... 21

1.3. Ion doping of semiconductors ... 27

1.4 Doping of semiconductors by nuclear transmutations ... 32

Chapter 2. Physical fundamentals of nuclear doping of semiconductors ... 38

2.1. Nuclear reactions as a source of impurity atoms ... 38

2.2. Nuclear reactions by charged particles ... 42

2.3. Nuclear reactions under the influence of γ-rays ... 48

2.4 Nuclear reactions by neutrons ... 51

2.5. Dopants and the nature of their distribution in semiconductors doped with (n, γ) reactions ... 57

2.6. The influence of side factors on nuclear doping ... 65

2.7 Possibility of obtaining p-n junctions by nuclear doping ... 75

Chapter 3 Radiation defects in semiconductors ... 84

3.1 Generation of simple radiation defects ... 84

3.2 Types of radiation defects in silicon ... 89

3.3 Accumulation of radiation defects ... 96

3.4 Interaction of radiation defects with various imperfections of the crystal lattice ... 102

3.5 Annealing of radiation defects ... 109

3.6 Radiation-accelerated diffusion ... 121

3.7 Radiation-induced defects in germanium ... 126

3.8. Radiation defects in binary compounds ... 128

Chapter 4 Silicon Nuclear Doping Technology ... 156

4.1. The technological scheme of the process of nuclear doping 156
4.2. Ensuring uniform irradiation of ingots in nuclear reactors 157
4.3 Specialized Nuclear Reactor 173
4.4. Annealing 178
4.5. Electrophysical properties of nuclear doped silicon 187
4.6. Effect of annealing medium on the properties of nuclear-doped silicon 198
4.7. The influence of the prehistory of the material and the parameters of the neutron flux on the electrophysical properties of nuclear doped silicon .. 211
4.7 Neutron flux density distribution in silicon ingots 216
4.8. Nuclear-Doped Silicon Radioactivity 219
4.9. The use of nuclear doped silicon in the manufacture of devices 223
4.10. Other applications of the nuclear transmutation method 225

«...The excellent uniformity of NTD silicon was finally demonstrated in 1971 by Kharchenko and Solov'ev who investigated radial spreading resistance for the first time (Izv. Akad. Nauk USSR Neorgan. Mat. 7, 2137 1971)). Their work was followed by extensive industrial research, which led to the first commercial introduction of NTD float zone for power rectifiers in 1974». (J. M. Meese, D. L Cowan and M. Chandrasekhar. Review of transmutation doping in silicon // IEEE Transactionson Nuclear Scienc. 1979. Vol. NS-26, № 6.)

Preface to the second edition

Since the first issue of the monograph, a rather long period has passed, which allows us to talk about the practical significance and prospects of doping semiconductors by the nuclear reaction method. Recently, the number of nuclear reactors has significantly increased, at which work was organized on the nuclear (transmutation) doping of semiconductor materials, in particular single-crystal silicon. According to the International Atomic Energy Agency (IAEA) [1], nuclear doping is carried out at least at 33 reactors located practically on all continents. It should be noted that in the past period, nuclear technologists managed to solve the problems of uniform exposure of large diameter silicon ingots to reactor neutrons. At present, doping of single-crystal ingots with a diameter of 150 and 200 mm, and in the future 300 mm, is most in demand. Moreover, the length of the ingots may be more than 500 mm. Work with such ingots requires the creation of appropriate technological equipment with a high degree of automation [1-5]. Note that the timely reaction of the "nuclear scientists" to the challenges of metallurgical methods of growing and doping large-sized ingots allowed them to survive in the tough competition of two fundamentally different doping methods. This is clearly evidenced by the dynamics of changes in the output of nuclear doped silicon (NDS) by years: in 1982, the total output was about 50 tons per year, by 1991 demand increased to 160 tons, then it began to fall and by 1999 reached 70 tons. However, after 2000, demand began to increase again and now it exceeds 100 tons per year. In other words, in the market of electronic devices in a certain niche, a noticeable amount of high-quality, uniformly doped silicon is required, which makes it possible to produce devices not only with the limiting electrical parameters, but also with a high degree of reliability.

In preparing the manuscript of the book, the authors took into account that a wide circle of specialists in various fields is involved in the problem of nuclear doping of silicon and its use in specific electronic devices: developers of electronic devices, metallurgists, specialists in nuclear and radiation physics, and material scientists. In this regard, the book includes four basic sections that complement each other or may be of independent interest.

The first chapter discusses various technological methods of doping bulk single-crystal ingots in the process of growing thin layers and (or) plates, as well as local regions by diffusion and ion implantation. Both the advantages and disadvantages of these doping technologies are discussed.

The second chapter provides basic information from nuclear physics on the interaction of neutrons, charged particles, and gamma rays with atomic nuclei of a target-semiconductor material. Using a number of semiconductors as an example, the effect of nuclear (transmutation) doping is distinguished and the corresponding experimental data are presented.

The third chapter discusses the formation and behavior of radiation defects in semiconductors that inevitably arise in the process of nuclear doping. The basic laws of the formation of the simplest point defects — vacancies and atoms displaced in the interstitial sites — as well as secondary, more complex formations such as vacancy clusters, defect complexes with impurity atoms, complexes involving interstitial atoms, disordered regions under the influence of neutrons, high-energy charged particles, and ions are given. The data on the interaction of radiation defects (primary and secondary) with various imperfections of the crystal lattice are discussed. An important element of nuclear doping technology — annealing of radiation defects — their transformation and annihilation is considered. The role of radiation-accelerated diffusion in the processes of formation and annealing of defects is noted.

The fourth chapter presents the main results of the practical implementation of a theoretically and experimentally sound technology for the nuclear doping of single-crystal silicon. In particular, the technique and technological methods of uniform irradiation by reactor neutrons of large ingots depending on the design features of nuclear reactors are considered. The problem of annealing of radiation defects is singled out separately. It is assumed that the completeness of annealing of defects can be judged by the data on the achievement of stable values of electrical resistivity, concentration and mobility of charge carriers. At the same time, the annealing modes vary depending on the irradiation conditions, the quality of the initial crystals, the sterility of the process, and other factors. Therefore, by agreement with the consumer, control tests of pilot batches are carried out, during which numerical values of the electrophysical properties specified in the contract (or in technical requirements), as well as permissible values of radioactive contamination are established (agreed) in accordance with sanitary standards.

Some data on other applications of the nuclear doping process are generalized. With cautious optimism, one can relate the prospect of nuclear (transmutation) doping to the possibility of growing single crystals of semiconductors from isotopically enriched semiconductor materials.

This edition of the book was undertaken in memory of my co-authors, well-known scientists who have done a lot for the development of radiation technologies in electronics.

V.A. Kharchenko

References:

1. Neutron Transmutation Doping of Silicon at Research Reactors / International Atomic Energy Agency. Vienna, 2012.
2. Neutron Transmutation Doping of Silicon. Massachusetts Institute of Technology. Nuclear Reactor Laboratory. URL: https://nrl.mit.edu/facilities/ntds.
3. Reactor Facility for Neu Transmutation Doping of Silicon Single Crystal. Institute of Atomic Energy POLATOM (IEA). URL: https://en.parp.gov.pl.
4. Carbonari A.W. Neutron Transmutation Doping of Silicon: Highli Homogeneous Resistivity Semicoductor Material in Nuclear Reactors/ IPEN-CNEN/SP https://www.ipen.br/biblioteca/cd/inac
5. S. Duun, A. Nielsen, Ch. Hendrichsen, Th. Sveigaard, O. Andersen, Ja. Jabłoński, L. Jensen. Neutron Transmutation Doping (NTD) Silicon for high Pover Electronics // TOPSIL. URL: www.topsil.com/media/56052.

Preface

Controlling the properties of semiconductors by doping them with the necessary impurities to specified concentrations is the main technological technique for creating any solid-state electronics devices. Impurity atoms are introduced into the energy gap of the semiconductor at local levels and serve either as electron suppliers or traps for them (donors or acceptors), or centers of radiative or non-radiative recombination of nonequilibrium charge carriers. It is controlled introduction of impurities that allows a wide variation of the conductivity of semiconductors, the type of conductivity, lifetime of nonequilibrium carriers, electron and hole mobility. Impurity atoms are introduced at different stages of the technological process into the bulk of ingots or layers during their growth, into thin layers so as to form any structures. Hence a fairly wide range of methods, the main ones are: introducing an impurity into a melt or a gaseous medium during the preparation of crystals and films, diffusing an impurity from surface sources, introducing an impurity from a beam of accelerated ions, forming an impurity subsystem due to nuclear transformations of atoms of the main or accompanying substance when irradiated by particles.

The rapid development of semiconductors lasts only the third decade, and if at first doping and diffusion methods satisfied the tasks, then with the development and complication of semiconductor devices, especially when switching to integrated circuits, as well as to large integrated circuits and high power power valves, serious limitations of traditional methods associated primarily with inhomogeneities of the material and genetic defects of the crystal, manifested during diffusion. The application and development of radiation methods, in particular, ion implantation, has opened up new prospects for the creation of ultra-complex systems of semiconductor electronics, when the sizes of individual active and passive elements began to be measured in microns (now nanometers). The successes of ion doping are due primarily to the extraordinary controllability of the process and its relatively low sensitivity to genetic defects. But even now, the initial inhomogeneities of crystals and layers are the main cause of marriage and the scatter of parameters of semiconductor devices.

The second, maybe even more important area of electronics is the production of high-voltage semiconductor valves, where super-rigid requirements for uniformity of properties on washers with diameters of about 100 mm are already being presented.

It was in connection with the solution of these problems that the attention of researchers was again drawn to the method of doping semiconductors using nuclear transmutations upon irradiation with γ-quantum, neutrons, or charged particles, since he promised to obtain uniformly doped crystals, even without microfluctuations, at great depths for neutrons and γ rays and uniformly doped in area for charged particles.

The idea of doping semiconductors using nuclear transmutations at the dawn of the semiconductor era was expressed by K. Lark-Horowitz. He first conducted experiments in Germany. Then it turned out undesirable moments that slowed down the development of the method - strong induced activity and, most importantly, significant changes in the properties of crystals due to the introduction of complex radiation defects.

At present, radiation defects have been well studied, ways to eliminate them have been found, silicon has become the main material of semiconductor technology. Its quality has increased, the content of uncontrolled impurities has fallen, and the relevance of obtaining uniformly doped materials has increased. That is why there is a sharp increase in interest in the nuclear doping of semiconductors, which was largely due to the fact that the method turned out to be successful for silicon. In this book - the first monograph on nuclear doping of semiconductors - an attempt is made to give complete information about this promising method.

The monograph combines issues from purely metallurgical to problems of applied nuclear physics and the technology of modern semiconductor electronics. We hope that it will be useful to a wide range of specialists.

The authors thank their comrades for useful advice and assistance in preparing the manuscript.

Please send criticisms of the book to the following addresses: 630090, Novosibirsk, 90, Prospect Nauki, 13, Institute of Semiconductor Physics; Obninsk, Kaluga Region, a branch of the Physico-Chemical Institute. L. Ya. Karpova.

L.S. Smirnov

Chapter 1. Doping of semiconductors

Doping of a semiconductor material is a controlled introduction of the necessary concentration of a certain impurity. At present, materials with different degrees of doping are used in the technique — from very "pure" crystals with an electrically active impurity concentration of less than 10^{12} cm^{-3} to heavily doped with an impurity concentration of more than $10^{20}cm^{-3}$. The planar technology for the manufacture of semiconductor devices imposes increasingly stringent requirements on doping technology. First of all, this is obtaining uniformly doped crystals and thin layers (1 - 10 microns) and doping of small volumes of complex configuration. For this reason, doping methods are being developed and improved. A detailed description of traditional methods and references to the original literature can be found in monographs [1-12].

1.1. Melt doping

From energy considerations, it follows that the impurity is not equally distributed between the liquid and solid phases. The distribution is characterized by a coefficient k, which is the ratio of concentrations in the solid (N_S) and liquid (N_L) phases. For equilibrium conditions:

$$k = N_S / N_L \tag{1.1}$$

For a two-component system consisting of a solvent and a dissolved substance, two qualitatively different situations should be considered: $k>1$ and $k<1$. For $k>1$, the melting point of the solvent with the introduction of an impurity increases, and for $k<<1$, the melting point goes down. For most impurities dissolved in germanium and silicon, the latter case is more characteristic. To illustrate the distribution of impurities between the two phases, small concentrations are usually taken, for which the liquidus and solidus lines can be considered straight. At higher concentrations, the lines bend and their position depends on the impurity concentration.

In the case $k>1$, the impurities are concentrated in the solid phase, which is enriched by them, and, conversely, when $k<1$, the impurities are driven away from the solid phase during solidification and the latter is purified. In both cases, concentration gradients arise in the liquid near the interface, and their relative steepness depends on the rate of impurity transfer in the liquid and the speed of the interface. For $k<1$, a layer enriched in impurity is created near the surface, and for $k>1$, a depleted impurity is created.

If the concentration in the melt is N_0, then the concentration of the impurity in the first portion of the solidified substance is kN_0. During further crystallization, the dissolved impurity atoms will concentrate on the partition surface, and as the volume of the liquid decreases, the concentration of the impurity in it will increase from the initial value N_0 to N_0k. The concentration of the impurity in the solid phase in this case increases from kN_0 to N_0.

The true segregation of impurities in a given crystal will be a function of the transport conditions prevailing in the liquid. These conditions can be classified as follows.

1. The impurity released at the interface is rapidly distributed in the melt due to strong mixing.
2. There is no convection by mixing in the melt, and the transfer of impurities from the interface is carried out exclusively by diffusion.
3. The intermediate case between conditions 1 and 2. The impurity is transferred through a thin surface layer of the melt δ into the volume by diffusion, and convection and mixing predominate in the volume. With regard to germanium and silicon, the latter situation is most frequent.

The first case is based on assumptions about the uniformity of concentration in the melt, the negligible smallness of diffusion in the solid phase, and the constancy of the value of k. The concentration of solute in liquid N_L can be written as

$$N_L = N(1-g) \tag{1.2}$$

where N is the amount of impurity remaining in the liquid after solidification of the volume g. Then, taking into account (1.1)

$$N_L = kN(1-g) \tag{1.3}$$

Based on differential equations (1.2) and (1.3) and taking into account that the initial impurity concentration is N_0, we obtain

$$N_S = kN_0(1-g)^{k-1} \tag{1.4}$$

where k - is the equilibrium distribution coefficient.

A completely homogeneous melt can be obtained by growing only under conditions of strong mixing. In many cases of crystal growth, the movement of impurities that are pushed away from the growth front occurs mainly due to diffusion, because technical difficulties cannot always be overcome.

The melt is partially mixed as a result of thermal convection, however, in the case of vertical growth, the temperature distribution may be such that the least dense part of the liquid is at the top. Therefore, under these conditions, natural convection is severely limited. With horizontal growth, a small cross section and a limited volume of the molten zone will also prevent convection flows.

The second case is analyzed on the basis of the following assumptions: mixing in the melt occurs only due to diffusion, convection flows can be neglected, diffusion in a solid is negligible, k remains constant.

Two main factors influence the distribution of an impurity in a liquid near a growing interface: a) the rate of diffusion of an impurity from an interface into a liquid, determined by the diffusion coefficient D (cm^2/sec); b) the rate of entry into the liquid of an impurity distilled off during solidification, determined by the velocity of the interface R (cm/sec) for $k < 1$.

Assuming that the interface is a plane perpendicular to the axis of the crystal, we can obtain a one-dimensional solution to the problem in the following form:

$$N_S = N_0\left[1-(1-k)\exp\left(-\frac{kR}{D}x\right)\right] \tag{1.5}$$

where x - is the distance from the interface.

The concentration of the impurity in the solid phase increases from the initial value kN_0 at $x=0$ to N_0, which corresponds to the average composition of the liquid. As the melt – crystal interface approaches the end of the sample, the equilibrium distribution of the impurity in the liquid shifts along the crystallization direction at a constant speed R. When the last portion of the liquid solidifies, its concentration becomes higher than the equilibrium value N_0/k and in the crystal it is higher N_0.

Crystals grown by the Czochralski method and the vertical zone melting method usually rotate with respect to the melt. As a result of this, the liquid is mixed, which greatly affects the distribution of impurities near the interface. Depending on the rotation speed, the fluid flow can be either

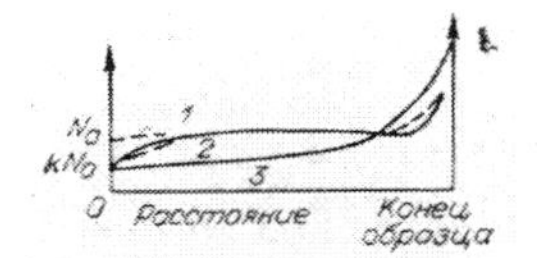

Figure 1.1.
Impurity distribution

turbulent or laminar, however, under any conditions there is a region near the interface in which the flow is laminar with a velocity decreasing to zero at the interface, i.e., near the interface there is a narrow liquid boundary layer in which the transfer of impurities is controlled by diffusion. It is assumed that, outside this boundary layer of width δ, mixing and convection flows provide a uniform concentration.

In the analysis of the third case, it is assumed that the surface has an impurity concentration gradient. According to this hypothesis, the concentration of impurities in the diffusion layer δ is greater than in the rest of the liquid, and determines the concentration in the solid phase. The thickness δ can be less than the thickness of the non-stationary liquid layer considered in the case of dominated diffusion (see paragraph 2), and therefore the concentration of the impurity at the interface does not reach the equilibrium value of N_0/k. For a finite growth rate, the effective distribution coefficient k_{ef} is calculated, which is the ratio of the concentration of the impurity in the solid phase on the N_S interface to the concentration in the liquid N_L outside the diffusion layer δ, i.e., $k_{ef}=N_S/N_L$, and its value is determined by the ratio

$$k_{ef} = \frac{k}{k+(1-k)\exp(-R\delta / D)} \tag{1.6}$$

where k - is the equilibrium distribution coefficient equal to N_S/N_0; $R\delta/D$ is a dimensionless quantity (including three main factors) that controls the value of k. The impurity distribution along the crystal, corresponding to three cases of theoretical approximation, is shown in the Figure 1.1. Curve 3 most closely corresponds to reality, since the value of the distribution coefficient k_{ef} lies somewhere between the equilibrium value and unity $(k<k_{ef}<1)$.

The concentration of the impurity in the liquid near the interface significantly affects the temperature gradient in the melt and, therefore, the behavior of the crystals during growth. Under certain conditions, the concentration gradient of the solute and the temperature gradient in the melt are such that conditions are created for the melt to be cooled. The liquidus temperature for a certain impurity concentration can be calculated from the phase diagram according to the following relation:

$$T_L = T_0 - mN_L \quad (1.7)$$

where T_L – temperature corresponding to the liquidus line for a liquid of a given composition; T_0 - melting point of pure solvent; m - slope of the liquidus line (assumed constant). Insofar as

$$N_L = N_0\left[1+\frac{1-k}{k}\exp\left(-\frac{R}{D}x\right)\right] \quad (1.8)$$

the liquidus curve equation is

$$T_L = T_0 - mN_0\left[1+\frac{1-k}{k}\exp\left(-\frac{R}{D}x\right)\right] \quad (1.9)$$

The true temperature in the melt in front of the interface at some point x is given by the equation

$$T = T_0 - mN_0 / k + G_L x \quad (1.10)$$

where T_0-mN_0/k interface temperature; G_L - temperature gradient in the melt. The conditions are possible when the temperature of the liquid is lower than its true solidification temperature, therefore, the liquid is supercooled. The condition for the absence of concentration hypothermia is determined by the ratio

$$G_L / R \geq mN_0(1-k)/kD \quad (1.11)$$

Usually, when doped semiconductors are grown, the impurity concentration is quite low and concentration overcooling is not observed. However, in the production of heavily alloyed materials, it can be an important factor. In this case, in order to obtain sufficiently perfect crystals with a uniform distribution of the impurity, it is necessary to grow them under conditions that allow avoiding concentration supercooling.

To obtain crystals with a uniform distribution of impurities, the zone melting method is used.

In the process of fractional crystallization, purification is easily carried out by alternating solidification and melting processes, if the impurity in the crystallization process is concentrated in a liquid $k<<1$. Before the last portion of the melt has solidified, the solid crystal is separated from the remainder of the liquid. Then the crystal is remelted again and, when solidified, is separated from the last portion of the melt. Since after each hardening, impurities are concentrated in the separated part of the liquid, the process can be repeated several times until the desired purification is obtained. In this case, part of the material is lost.

In the zone cleaning method, the liquid zone can pass through the ingot several times, without significant loss of material, except for evaporation. The process can be significantly accelerated by using molten zones moving sequentially one after another.

Purification of Germany in the form of a bar located in a graphite boat is carried out without much difficulty. Purification of silicon due to its high chemical activity in the liquid phase is more complicated. It mainly uses the method of vertical, or crucible-free, zone melting. The use of the latter method is limited because the process cannot be accelerated by the use of simultaneously passing liquid zones.

It was experimentally found that after the first passage of the molten zone along the ingot, three regions are observed that differ in the nature of the distribution of the impurity. The concentration at the beginning of the ingot increases, reaches a constant value, and at the end it rises steeply again. The shape of the impurity distribution curve, among other factors, depends on the value of k. During initial melting, the liquid phase has an initial concentration of N_0. As it moves forward, the first portion of the resulting solid phase has a concentration kN_0, which is less than N_0 for the case $k<1$. As a result, the liquid phase is continuously enriched in the impurity. The concentration of solute in the solid phase also increases from the initial value of kN_0 at $x=0$. When an equilibrium occurs between the impurity entering the liquid phase and being removed from it, the concentration of the impurity in the solid phase will be constant. When the end part of the ingot solidifies, the zone length decreases to zero and the impurity concentration sharply increases.

When solving the simplest equation that determines the distribution of impurities along a solid ingot after the first passage of the zone, the following expression is obtained for N_S

$$\begin{gathered} N_S = N_0(k-1)\exp(-kx/l) + N_0, \quad or \\ N_S / N_0 = 1-(1-k)\exp(-kx/l) \end{gathered} \tag{1.12}$$

where l - zone width.

The degree of purification with one pass of the zone is less than with conventional solidification. Moreover, for $k<0.5$, the initial transitional distribution region extends over the entire length of the ingot, i.e., a horizontal section with a constant concentration is not observed.

For $k<1$, the zone must travel a considerable distance along the ingot before the amount of impurities accumulated in the melt is sufficient for the equilibrium concentration N_0/k, at which the amount of impurity entering and leaving the liquid zone remains constant. If the concentration of the impurity in the liquid zone at the beginning of its occurrence is N_0/k, then the initial transition region in the distribution curve can be eliminated.

Therefore, by adding the required amount of the desired dopant to the liquid phase at the beginning of its passage along a clean ingot with a uniform cross section, a constant distribution of the impurity can be obtained. After a single passage through the zone, the N_S concentration is represented as

$$N_S = kN_{in} \exp\left(-kx / l\right) \tag{1.13}$$

where N_{in} - initial concentration of impurities in the liquid zone.

Significantly improve the distribution of impurities can also be by changing the direction of passage of the zone. Deviations after the first pass will be partially compensated during the return pass, and with sufficient alternation of the forward and reverse passages, a fairly uniform distribution of the impurity along the entire ingot will be obtained, with the exception of, of course, the last solidified zone.

Other methods for producing crystals with a uniform distribution of the impurity are reduced to adding the desired impurity to a very clean melt at a rate at which it is removed from the hardening surface.

One of the methods for obtaining a uniform distribution of impurities is to control the partial vapor pressure of the alloying element. Since the concentration of the impurity in the liquid phase increases with increasing distance from the beginning of the ingot, the partial pressure of the vapor above the liquid increases. Therefore, while maintaining the desired partial pressure of the impurity, it is theoretically possible to maintain a constant concentration of the impurity in the melt. A modification of this method has been successfully used in growing crystals under reduced ambient gas pressure. The loss of impurities due to evaporation from the melt can be balanced by the release of impurities from the growing crystal $k<1$.

Another variant of the method, which allows maintaining a constant concentration of impurities in the melt, is a constant replenishment of the melt. The polycrystalline rod is continuously fed into the melt at a rate exactly equal to the drawing speed of the single crystal, i.e., the melting and solidification rates are strictly controlled so that the volume of liquid material in the crucible remains constant. Feeding the melt with an ingot with a given impurity concentration, it is possible to draw a crystal with the desired concentration, while the impurity concentration in the liquid also remains constant.

As already noted, the distribution of impurities in a crystal when it is drawn at a constant speed from the melt varies exponentially along its length. Since the effective distribution coefficient k_{ef} varies with the growth rate to the same extent as with the rotation speed, a change in the content of impurities can be partially compensated by monitoring these parameters. For example, the value $k_{ef}<<1$ increases with increasing growth rate, approaching unity for very high speeds, that is, the impurity concentration increases. Therefore, by programming the growth rate (drawing speed), it is possible to obtain a uniform distribution of the impurity along a considerable length of the ingot. In practice, crystals are rapidly elongated at the initial stage, and as the impurity concentration in the melt increases, the extrusion rate slows down to the desired value. The disadvantage of this method is that the drawing speed can be changed only in a relatively limited area, consistent with the principles of growing perfect single crystals.

The uneven distribution of impurities in a single crystal can be due to the action of both random and regular factors. Random are caused by uncontrolled changes in the crystallization process due to imperfections in the equipment for growing single crystals. These include: power fluctuations in the heater circuit in the absence of stabilization of the thermal regime of the installation. asymmetry of the thermal field in the melt at the phase boundary, fluctuations in the growth rate due to poor manufacturing quality of mechanical drives and the lack of stabilization of their power supply, etc.

The action of natural factors is due to the peculiarities of the process of growing single crystals: natural segregation of the impurity along the length of the grown ingot, uneven distribution of the impurity in the volume, due to the periodic nature of the crystallization process, the manifestation of various crystallographic faces during growth, the shape of the crystallization front, nature and the concentration of the impurity itself, the interaction of the dopant and other impurities, dislocations, etc. Inhomogeneities of this type can be studied and minimized by selecting appropriate growing conditions.

Detailed studies have shown that the distribution of impurities along the length of the grown crystal under real conditions is far from a smooth curve and is periodic. Periodic heterogeneity of the distribution of impurities in the growth bands is a serious source of volume heterogeneity. Periodic fluctuations in the content of impurities were observed both in crystals grown by the Czochralski method and in crystals obtained by horizontal band or directional crystallization. In elongated single crystals, several systems of growth bands were found; in addition to periodic inhomogeneities associated with the rotation of the crystal and crucible, small growth bands with a period of 3–20 μm were also observed. The latter are also found in crystals grown without rotation. The periodic inhomogeneity associated with the rotation of the crystal and the crucible is apparently caused by periodic fluctuations in the growth rate due to the presence of an asymmetric thermal field in the melt at the interface. A more complex effect is the small growth bands observed in crystals obtained by different methods. The reasons for their occurrence lie, apparently, in the periodic nature of the crystallization process itself, one of the external manifestations of which can be found in practice, self-oscillations of phase separation. In some works, the occurrence of self-oscillations of the crystallization front is considered from the standpoint of deviation from thermodynamic equilibrium at the phase boundary.

In other works, the appearance of periodic inhomogeneities is associated with the presence of concentration supercooling in the melt. But growth bands are also found in very pure single crystals, where there is no concentration supercooling. Nevertheless, the dependence of periodic inhomogeneity observed in heavily doped single crystals on the nature of the doping impurity, and primarily on the magnitude of the distribution coefficient, indicates that the accumulation of impurities in the diffusion layer of the melt at the crystallization front plays a role in the formation of the growth band.

One of the sources of the uneven distribution of impurities in the volume of a single crystal is the dependence of the effective distribution coefficient on the orientation of the surface of the crystallization front.

The introduction of most impurities into the diamond lattice (in the nodes) causes a significant distortion of interatomic bonds. Therefore, from energy considerations, it can be expected that their solubility should be generally small. This is true when the difference in atomic sizes of the solute and solvent is significant. It was shown that with a difference in atomic radii of more than 15%, the atomic size factor is unfavorable and low solubility should be expected. Therefore, comparing the ionic radii of impurity elements, we can tentatively assess their solubility in germanium and silicon.

In addition to the geometric dimensions of atoms (ions), the solubility of the impurity is also affected by electrochemical factors. Atoms in diamond-type lattices are held in a tetrahedral arrangement by four covalent bonds of the SP^3 type. Elements III - V of the group have valence electrons in the *S*- and *P*- states and, therefore, as a result of hybridization, the orbits of these electrons can easily participate in the SP^3 bonds. This type of connection is most pronounced for elements of group IV. The highest solubility is characteristic of electrically active impurities of elements of groups III and V. The dissolution of transition elements (iron, nickel, cobalt) and elements of the IB group (copper, silver, gold) in the diamond lattice causes relatively strong electrical effects and distortions, so their solubility is small. In addition to these factors, when evaluating the solubility of one element in another, their electronegativity should be taken into account.

As already indicated, depending on the distribution coefficient k, the impurity will be concentrated in the liquid or solid phase. It is necessary to know k different electrically active impurities to compare their relative tendency to dissolve in solid material. The values of k can be obtained directly from the phase equilibrium diagram, which indicates the chemical composition of phases in equilibrium at various temperatures. The diagrams are constructed for concentrations of solute of not less than 1%. For semiconductors, much lower concentrations have to be considered. By extrapolating to such low concentrations using conventional phase equilibrium diagrams, only approximate values can be obtained. Often, the phase diagram for a two-component system is complex and you have to enter two distribution coefficients (for example, a germanium – antimony system).

In a simple binary system, the direction of the solidus line in the phase diagram is usually monotonous from the melting point of this component to the eutectic line. However, in many cases, the solidus line reveals an inflection before the eutectic line is reached; in this case, the concentration of the impurity in the solid phase initially increases to a maximum and then decreases with a further decrease in temperature. This behavior is called retrograde solubility.

The expressions for the distribution coefficients given above are obtained from theoretical considerations. Although these values are in reasonable agreement with experimental data, they are nevertheless approximate.

Based on thermodynamic considerations, the distribution coefficient can be expressed as a function of absolute temperature:

$$\mathrm{In}k = (T_{me} / T)\,\mathrm{In}k_m + (\Delta S^f / R)(T_{me} / T - 1) \tag{1.14}$$

where k_m — constant value taken at the melting point (T_{me}) pure solvent; ΔS^f - entropy of melting of an impurity at its melting temperature; R - gas constant.

The dependence of the logarithm of the distribution coefficient, expressed by equation (1.14), on the inverse absolute temperature has the form of a straight line. Extrapolating it to T_{me} gives the value k_m. In most cases, the k_m value obtained in this way agrees well with the experimental data.

In Germany, a correlation was found between solubility in the solid state and atomic size. For germanium and silicon, a correlation was established between the distribution coefficients of various impurity elements and the corresponding heats of sublimation. A simple correlation was observed between the maximum molar solubility of X impurity elements in germanium and silicon and their distribution coefficient k_m at the melting point:

$$X = 0,1k_m \tag{1.15}$$

The solubility of impurities can be controlled by external exposure. For example, according to the law of masses, an increase in electron concentration decreases the solubility of donors.

The solubility of this impurity element is affected by the presence of another donor or acceptor impurity.

1.2. Impurity diffusion

Impurity diffusion is one of the main technological methods in the manufacture of semiconductor devices. Despite the fact that this method has been used in technology for many years, a positive result is achieved empirically - by trial and error. Issues related to the diffusion of impurities in semiconductors are covered in sufficient detail in monographs [2, 3, 6, 7, 12]. They summarized only some issues of diffusion in semiconductors as a method of introducing an impurity and attempted to emphasize those factors that influence the deviation of the actual distribution of an impurity from theoretical, calculated.

The diffusion process of impurity atoms in interstitial solid solutions is the simplest and most obvious example of diffusion in solids. The diffusion mechanism here is reduced to a sequential transition of atoms from one interstitial site to another. In the case of dilute solutions, one can consider the hopping of an atom from one interstitial to another as independent of each other and consider diffusion as a process of random motion of particles.

The diffusion mechanism of dissolved atoms in substitutional solid solutions is more complex than diffusion in interstitial solid solutions. Dissolved atoms can move in the crystal in various ways: by hopping to vacant sites, through interstitial sites, by direct exchange of atoms by places, by a ring mechanism. The question of which of these mechanisms is actually being implemented requires additional research in each case. However, the vacant diffusion mechanism is most frequently encountered.

In the case when a vacancy forms near a dissolved atom more easily than in other places of the crystal, diffusion can occur by moving complexes consisting of diffusing atoms and vacancies. The process of hetero diffusion will also be accompanied by the process of diffusion itself, however, vacancies due to lower binding energy exchange places with foreign atoms much faster than with solvent atoms.

Additional questions about diffusion mechanisms arise when considering the interaction of diffusing radiation defects with impurities and other imperfections of the crystal lattice. There is no consensus on the mechanism of radiation-accelerated diffusion of an impurity. The book [12] describes diffusion, which takes into account the experimental data obtained in radiation physics. It is assumed that in crystals the migration energy is low not only for vacancies and interstitial atoms, but also for other point defects, in particular impurities, but the migration process for them is complicated by "sticking" or complexation. When considering diffusion, one proceeds from the following provisions [12]: 1) the migration energy of vacancies and interstitial atoms is small. Migrating along the crystal, vacancies and interstitial atoms interact with impurity atoms and other defects, forming complexes. Annihilation of vacancies and interstitial atoms and their occurrence is a special case of the formation and decay of complexes; 2) the formation and decomposition of the complexes are determined by the migration energy of the moving particle, the complexation barrier and the binding energy of the complex; 3) all energy characteristics depend on the charge state of the reacting defects; 4) the impurity migrates either in the form of a complex with a vacancy, or, after being displaced from the site, at interstitial positions until it is captured by any defect. The movement of vacancies is limited by their capture and subsequent release from the trap. Therefore, the non-identity of the experimentally determined diffusion parameters is explained by the difference in the energy characteristics of the complexes and the difference in the types and concentrations of traps for diffusing particles.

Whatever the mechanism of the process, atoms can only move through the lattice when they have thermal energy sufficient to overcome the energy barrier surrounding their equilibrium positions, i.e., to jump from one position to the next. The diffusion process in this case can be considered as a purely statistical decrease in the concentration gradient in the crystal as a result of random jumps.

The relationship between the flow velocity of the diffusing impurity and the concentration gradient causing this flow in a certain x direction is expressed by the first Fick law of diffusion

$$J = -D(dN / dx) \tag{1.16}$$

where J - flux density, i.e., the number of atoms diffusing through a unit area per unit time; N - concentration of diffusing impurities in the x direction; D - diffusion coefficient. This equation describes a concentration drop that is proportional to its gradient. Under real conditions of diffusion, the concentration gradient varies over time. Temporal dependence is described by Fick's second law.

Diffusion coefficient D is expressed as a function of temperature relation

$$\frac{\partial N}{\partial t} = D\frac{\partial^2 N}{\partial x^2} \tag{1.17}$$

$$D = D_0 \exp(-E / kT) \tag{1.18}$$

where E - diffusion activation energy; D_0 - diffusion constant; k - Boltzmann constant. Equation (1.17) is valid if the diffusion coefficient is independent of the concentration of diffusing particles.

We give solutions of the diffusion equation for some special cases.

a) Infinitely thin diffusion

$$N(x,t) = \frac{Q}{2\sqrt{\pi D t}} \exp\left(-\frac{x^2}{4Dt}\right) \tag{1.19}$$

where $Q = N_0 \times 2h$ - impurity concentration N_0 in a thin layer thick 2h. Relation $N(x,t)$ from x given by the factor $exp(-x^2/4Dt)$, therefore, an infinitely thin layer at any time gives a symmetric distribution of the impurity concentration. The concentration reaches its maximum value at $x=0$. The maximum value decreases with time:

$$N_{\max} = Q / 2\sqrt{\pi D t} \tag{1.20}$$

b) Diffusion from a layer of finite thickness

$$N(x,t) = \frac{N_0}{2}\left[\operatorname{erf}\left(\frac{h+x}{2\sqrt{Dt}}\right) + \operatorname{erf}\left(\frac{h-x}{2\sqrt{Dt}}\right)\right] \tag{1.21}$$

where is the error function

$$\operatorname{erf}(y) = \frac{2}{\sqrt{\pi}} \int_0^y e^{-z^2} dz$$

N_o - concentration at the initial time in the field ($-h$, $+h$). As $h \to 0$, expression (1.21) goes over into (1.19).

c) Diffusion from semi-infinite space. The initial distribution is such that for all $x<0$ $N(x, 0)=N_0$, but for everyone $x>0$ $N(x, 0)=0$. The solution (1.17) is obtained in the form

$$N(x,t) = \frac{N_0}{2}\left[1\text{-}\operatorname{erf}\left(\frac{x}{2\sqrt{Dt}}\right)\right] \tag{1.22}$$

It is noted that for all $t>0$, the concentration in the interface ($x=0$) is constant and half N_0.

The solutions of equation (1.17) for a semi-bounded body under various boundary conditions are given in the monograph [2].

Until now, when considering diffusion, the electric charges of diffusing particles and the resulting interaction between them were not taken into account. Such an interaction between various impurities, as well as between impurities and crystal lattice defects (vacancies), can significantly affect the impurity diffusion in semiconductors. Of particular interest is the case when the diffusing impurities have an opposite electric charge in sign, that is, when the diffusion of the donor impurity and the acceptor impurity occurs simultaneously. The Coulomb interaction between such ions can lead to the formation of an association, which reduces the diffusion rate and affects the processes associated with the scattering of mobile charge carriers in semiconductors. The theory of the processes of interaction and complexation during the diffusion of impurities in semiconductors was considered by Rice, Fuller and Murin. They examined a hole-type semiconductor in which donor diffusion takes place: N_a - concentration of acceptor impurity uniformly distributed throughout the volume, $N_d(x)$ - donor concentration at point x, $P(x)$ - concentration of pairwise bound ions of donors and acceptors (complexes). Free donor concentration $(N_d - P)$. Fick's first law can be written as follows

$$J = -D_0 \frac{\partial (N_d - P)}{\partial x}$$

where D_0 - diffusion coefficient of free donors (in the absence of complexation). Using the law of mass action, we obtain $J=-D(dN_d/dx)$. Here we get:

$$D = \frac{D_0}{2}\left[1 + \frac{\frac{1}{2}\left(N_d - N_a - \frac{1}{k(T)}\right)}{\sqrt{\frac{1}{4}\left(N_d - N_a - \frac{1}{k(T)}\right)^2 + \frac{N_d}{k(T)}}}\right] \quad (1.23)$$

where $k(T)$ – equilibrium constant in the law of the acting masses.

Thus, in the case of the formation of coupled complexes between donors and acceptors, the diffusion coefficient is a function of the concentration of donor and acceptor impurities, as well as the equilibrium constant.

When solving diffusion equations with allowance for complex formation, it was assumed that the diffusion coefficient of the complexes is negligible and the diffusion of the complexes was not taken into account. But there are cases when the diffusion coefficient of the complex significantly exceeds the diffusion coefficient of the free impurity.

The diffusion of impurity ions in semiconductors is often superimposed by the ion drift in the electric field, which arises due to local electrostatic diffusion potentials. Such a field exists, for example, in electron-hole transitions, as well as in areas of non-uniform distribution of impurity ions.

An internal (gradient) electric field in the simplest case arises when an impurity diffuses into a homogeneous semiconductor. The reason for the appearance of this field is the ambipolarity of diffusion and the difference in the diffusion coefficients of moving ions and their accompanying charge carriers (electrons or holes). Charge carriers have a much larger diffusion coefficient and the opposite sign of electric charge with respect to diffusing ions. In their movement, they are ahead of ions. This leads to the appearance of an electric field, which is directed in such a way that accelerates the movement of ions and slows down the movement of charge carriers.

Along with the vacancy and interstitial diffusion mechanisms in semiconductors, simultaneous diffusion of an impurity over vacancies and interstitials is often observed. In this case, both in vacancies and in interstitial particles, impurity particles move in different charge states, which leads to the appearance of several interconnected diffusion flows between which particles are continuously exchanged. Such a diffusion mechanism is called dissociative. An exact mathematical description of this diffusion mechanism is rather complicated and in the literature there are only approximate solutions for specific cases (for example, copper in germanium, zinc in gallium arsenide).

In the presence of a constant external electric field, an additional force acts on the impurity particle due to the scattering of ions of electrons and holes moving to the electrodes. This force carries impurity particles in the direction of motion of electrons and holes. The effect can significantly change the mobility of the ion and even the direction of its movement to the opposite. So, by applying an electric field, you can to some extent affect the diffusion process.

At high concentrations of diffusing impurities, the dependence of the diffusion coefficient on the concentration of the impurity and the interaction between the diffusing particles are manifested. With a high level of doping, the dopant exists in the crystal simultaneously in several forms, and not all of them are electrically active. An analysis of the data for germanium and silicon shows that, in addition to the usual dopant atoms, there are complexes including several dopant atoms. So, for the Ge-Sb, Ge-As, Si-As systems, the 4 Sb and 4 As complexes were detected [7]. It was also established that in the same systems the formation of a phase containing up to 40-50% of the dopant is possible. If the crystal in which diffusion occurs contains dislocations or other large structural defects, then decoration with a diffusing impurity of these disturbances is observed.

All the data presented indicate that, despite the great efforts of many experimenters and theorists, it is not always possible to predict the exact result of diffusion. In addition, purely technical difficulties arise when doping thin washers of large diameter and when producing very thin doped layers (~ 1 μm). The first difficulty is due to the fact that diffusion of the impurity usually occurs at high temperatures (1100–1350° C) and for a long time, which leads to deformation of the washer; the second is the impossibility of reproducibly diffusing to very shallow depths.

1.3. Ion doping of semiconductors

Ion doping is the introduction of atoms into the surface layer of a substrate material by bombarding it with ions with energies from a few kiloelectron-volts to several mega-electron-volts. The ion doping method is very promising for semiconductor electronics, and in some cases has obvious advantages over traditional doping methods (diffusion, melt doping). The advantages of this method are mainly associated with the non-thermal nature of the doping. These include: 1) universality, that is, the possibility of introducing any impurity into any solid substance; 2) low temperature - usually concomitant annealing is carried out at temperatures significantly lower than with diffusion doping, which avoids undesirable high-temperature effects on the semiconductor material (for example, distortion of the plane of the alloying front); 3) the ability to flexibly control the distribution of impurities in all three dimensions by varying the ion energy, using protective masks and scanning the beam; 4) the possibility of a strict dosage of impurities during alloying with the help of precise control of the ion current density and irradiation time; 5) the purity of the introduced impurity, which is ensured by the electromagnetic separation of ion beams and the vacuum process conditions; 6) the possibility of doping through dielectric and metal coatings.

At the same time, the ion doping method has some limitations and disadvantages: 1) the small depth of ion penetration; 2) the presence of radiation defects; 3) insufficient knowledge of physical processes during ion doping; 4) the relative complexity of the equipment at this stage of the development of the method.

Given all the disadvantages and advantages of the method, it can be argued that it does not deny other doping methods, on the contrary, in combination with them it becomes one of the most important components of semiconductor technology.

The main factors that determine the effectiveness of the practical use of the ion doping method are the mean free path distribution of the embedded atoms, the degree and nature of the disorder of the lattice created upon implantation, the localization of the embedded atoms in the crystal lattice, and the electrical characteristics after ion implantation and subsequent annealing.

Any analysis of processes during ion doping begins with the question of the distribution of embedded atoms in depth. The distribution of the ranges of ions depends on the processes of their braking. Recently, many theoretical and experimental works have appeared in which the processes of ion deceleration have been studied, and now it is possible to calculate the distribution parameters with reasonable accuracy. In the amorphous substrate, the range distribution is approximately Gaussian; it can be characterized by the average range and spread relative to this average position. The average range depends on the energy of the bombarding particle, and the spread is significantly affected by the mass ratio of the incident ion and matrix atoms. A feature of ion doping is the ability to obtain, in principle, any impurity distribution profile, changing the energy of the bombarding ions. The average range of ions with an energy of several tens or hundreds of kiloelectron-volts is about 0.1 microns. This ionic doping differs from diffusion, with which it is difficult to obtain doped layers with a thickness of less than 1 micron.

In a crystalline substrate, as shown in numerous experiments, the range distribution strongly depends on the orientation of the crystal with respect to the direction of motion of the ion — the channeling effect. If an ion enters the crystal almost parallel to one of the principal axes or planes, then it can move in the lattice, experiencing only weak collisions at a sliding angle. With this motion, the rate of energy loss decreases and the ion penetration depth increases. It was experimentally found that channeled ions penetrate to depths that are almost an order of magnitude greater than the average range. It is clear that the channeling effect leads to a change in the distribution profile of the impurity when compared with the profile calculated for an amorphous target. The channeling process depends on many factors, sometimes difficult to control, so it is difficult to predict the shape of the distribution of impurities in the presence of channeling. It was experimentally shown that only a small part of the impurity remains in the channeling mode or enters into it. The most important factors on which the distribution of channeled ions depends are as follows: 1) temperature of the target (substrate); 2) dose of ions; 3) misorientation and disordering of the surface.

Sometimes the profile of the distribution of impurities can be affected by the effects of accelerated diffusion. One of such mechanisms is the radiation acceleration of diffusion by increasing the concentration of vacancies. This effect is noticeable at temperatures several hundred degrees below those at which conventional diffusion is carried out. Another mechanism is associated with the diffusion of atoms over interstitial sites and is effective at even lower temperatures. In this case, the concentration of defect centers is significant, which can serve as traps for interstitial atoms (for more details, see Chapter 3).

A number of problems arising from the use of ionic doping are related to the disordering of the lattice and radiation defects created by the implanting ion. In the process of deceleration, the ion collides with the atoms of the lattice, knocking them out of the nodes. Some knocked out atoms have such energy that they themselves can displace other atoms, resulting in a disordered region. At high radiation doses, individual disordered regions overlap, forming a continuous disordered, sometimes amorphous layer. In addition, many simpler defects arise in the irradiated layer, including vacancies, interstitial atoms, and various impurities (including dopants). To obtain the doping effect, annealing of the introduced defects is fully necessary.

Isolated disordered regions and an amorphous layer behave differently during annealing. In Germany, disordered regions are annealed at a temperature of 200°C, and in silicon at 300°C, while an amorphous annealing layer requires high temperatures: 600°C for silicon and 400°C for germanium. Point defects and complexes are annealed, as a rule, at lower temperatures. However, even after 600-degree (for silicon) annealing, not all defects are completely annealed. This is evidenced by an increase in the number of dislocations in silicon at annealing temperatures above 600°C. As the experiment shows, after annealing at 700-800°C, the properties of ion-doped layers are close to the properties of volume-doped material.

An impurity is electrically active if it is in a certain state. For example, an aluminum atom located in a silicon lattice site is an acceptor, while in the interstitial position it is a donor. An admixture that decorates dislocations or other major imperfections does not exhibit electrical activity. This example is enough to understand how important it is to know where the impurity is located. Currently, the concentration of electrically active impurities is determined from measurements of the Hall effect, and the impurity position in the crystal lattice can be determined by the channeling effect of light particles. For some impurities, a correlation was established between annealing of defects, an increase in the number of impurities in the sites, and the concentration of charge carriers in the ion-doped layer. It is most clearly found in silicon doped with elements of group V at room temperature. It has been experimentally shown that the concentration of substitute antimony and the concentration of electrons during annealing change identically. In the case of irradiation, when ions were embedded in a heated substrate (substrate temperature above 300°C), there was no direct relationship between the concentration of nodal antimony and the concentration of electrons: immediately after irradiation, almost all antimony was in the sites, but the electron concentration approached the concentration of antimony only after annealing at $T>700°C$. The reason for this discrepancy is the presence of defects acting as compensating centers. The situation is much more complicated when doped with elements of group III. Here, substitutional and interstitial atoms are often found simultaneously. The electrical properties of group III atoms in silicon also turn out to be complex; here, annealing temperatures of more than 800°C are required to exhibit the maximum electrical activity of the impurity.

We present the data on the localization of various impurity atoms known for silicon. By measuring the channeling effect in the <111> and <110> directions, as well as the output for a non-channelized beam, one can determine: 1) the fraction of embedded atoms located at a distance of 0.1-0.2 Å from the lattice sites (S); 2) the proportion of embedded atoms located at a distance of 0.1-0.2 Å from the tetrahedral positions (I); 3) the fraction of randomly located embedded atoms (R) that do not occupy positions 1 and 2. If the number of point defects is not too large, then this kind of measurement allows us to distinguish four cases:

I. $S \neq 0$, $I \approx 0$, those, a significant proportion of interstitial atoms is located at the lattice sites and there are no interstitial atoms. In silicon (and germanium), usually 50-90% of group IV and V impurities are present at substitution sites.

II. $S \approx 0$, $I \neq 0$, those, a significant proportion of interstitial atoms is in interstitial positions and there are no substitutional impurity atoms. This situation is quite rare, but it can be observed under special conditions for the introduction of thallium and elements of group II. Usually $I \leq 0.6$.

III. $S \neq 0$, $I \neq 0$, i.e., implanted atoms are located both in lattice sites and in interstitial positions. This situation is usually found for elements of group III in silicon and germanium, but has never been observed for elements of group V. Often $S \approx I \approx 0{,}3$.

IV. $S \approx I \approx 0$, $R \approx 1$ those, and in the nodes and in the interstitial positions of the implanted atoms are few. This is usually observed for elements of group III upon high-temperature irradiation or after annealing at high temperature. It is usually believed that the R component is associated with atoms that have gone to dislocations or other imperfections, or are enclosed in defective regions. But another interpretation is also possible: 1) the atoms occupy not tetrahedral interstices, but, say, hexagonal ones; 2) the atoms are displaced from the position corresponding to the site or tetrahedral interstitial by more than 0.2 Å. In the latter case, the measured values of S and I will be underestimated.

The localization of an atom is significantly affected by the interstitial temperature and subsequent annealing, the dose of ions, and the type of impurity.

For ion doping, large concentrations of the introduced impurity are characteristic. For example, it presents certain difficulties to introduce impurities in an amount of less than 10^{11}-10^{12} cm^{-2}, and with a range of 10^{-5} cm this corresponds to a concentration of 10^{16} - 10^{17} cm^{-3}. The experiment shows that during ion doping, impurity concentrations are easily reached that exceed the solubility limit of this impurity.

1.4 Doping of semiconductors by nuclear transmutations

By traditional technological methods, it is increasingly difficult to satisfy the requirements of uniformity and accuracy of doping. In this regard, in recent years more and more attention has been paid to nuclear, or transmutation, doping of semiconductors. The fundamental possibility of transmutation of nuclei under the influence of γ-radiation, neutrons, protons, deuterons, α-particles has been known for more than half a century, but the use of this method has long been constrained by insufficient knowledge of radiation processes and the behavior of impurity atoms. The first experiments on nuclear doping of semiconductor material were carried out in Germany. Then came the turn of silicon and binary compounds. This method is widely used for doping silicon. In this case, as a rule, reactions of silicon atoms with neutrons are used.

When neutrons interact with silicon isotopes, excited composite nuclei are formed. For a composite nucleus, there are several energetically possible decay paths, for example: γ-ray emission (reaction (n, γ), neutron scattering (n, n), emission of one or more particles (reactions $(n, 2n)$, (n, p), (n, α). These processes compete with each other, and under certain conditions there is only one possibility.

In silicon, under the influence of thermal neutrons, the reaction (n, γ) takes place, leading to the formation of a dopant - phosphorus (this issue is described in detail in Chapter 2). The reactions accompanied by the emission of n, p, α particles (n is a neutron, p-proton, α -α particles) are endothermic and occur only under the influence of neutrons with sufficiently large energy (see Chapter 2).

Let us dwell on some side effects observed during transmutation doping of a semiconductor material. This is primarily artificial, or induced, radioactivity. Distinguish surface activity and sample volume. Surface activity is composed of the activity of impurities that were on the surface of the sample, and the products of corrosion and sputtering of the reactor materials (when using neutrons). Volume activity is caused by nuclear reactions with existing impurities and matrix atoms. Surface activity is disposed of by mechanical or chemical cleaning of the surface of the ingot.

Volumetric activity depends on the half-life of active isotopes. For example, the half-life of the Si^{31} isotope ($Si^{31} \rightarrow P^{31}$ transmutation) is 2.6 hours (β^- decay), and the activity drops to a safe level in a few days. At the same time, germanium isotopes have a half-life of 76 days, and it takes many months to wait until the radioactivity drops to a safe level. Already from these examples it is clear that with transmutation doping, the use of elements with long-lived radioactive isotopes should be avoided.

Nuclear reactions are accompanied by the emission of γ-quantums, β-particles (electrons, positrons) and heavier particles, such as neutrons, protons, α-particles, etc. Having a high energy (about 10^6 eV), these particles and quanta lead to the appearance of radiation defects in the volume of the material. The types of radiation defects arising under the influence of γ quantums, electrons, and protons are considered in Chapter 3.

When γ-quantum are emitted, as follows from the laws of conservation of energy and momentum, the nucleus receives significant energy. As calculations [13] showed, the nucleus of a silicon atom acquires an energy of 780 eV on average, and the maximum recoil energy reaches 2010 eV. This excess energy is sufficient to displace an atom from a site and displacements of nearby atoms, i.e., to form a disordered region.

In nuclear reactions accompanied by the emission of particles, a kind of internal irradiation of the crystal by these particles occurs, and radiation defects similar to those formed during ion doping occur.

During transmutation doping, the material is usually placed in a nuclear reactor, which has a fairly wide range of neutrons in energy (see chapters 2 and 4). During scattering, fast neutrons transfer enough energy to matrix atoms to form a cascade of displacements or a disordered region (in addition to the possibility of transmutation reactions). Similar effects are observed if transmutation reactions proceed under the influence not of neutrons, but of charged particles.

Thus, during transmutation doping, radiation defects are formed under the influence of γ radiation, electrons, protons, α particles, recoil nuclei and, in some cases, nuclear fission fragments, which leads to the accumulation of a spectrum of defects from simples (such as vacancy + impurity complexes, divacancy) to disordered areas.

The presence of radiation defects masks the effect of doping immediately after irradiation under conditions that ensure nuclear transmutation; therefore, annealing of radiation defects is a necessary technological step. The annealing mode (primarily temperature and duration) must be chosen so as to eliminate the adverse effect of introduced defects. The data available in the literature show that the optimal annealing mode has not yet been selected (see also chap. 4). This is due to the following. Transmutation doping itself is a method that does not depend on the temperature of its implementation, and the most high-temperature operation is annealing of radiation defects. There is no clearly defined annealing criterion. So, to achieve the calculated concentration of phosphorus in silicon, annealing for 0.5–2 h at 800–850°C is sufficient; i.e., under these conditions, compensating conductivity associations are annealed. However, other research methods (neutron diffraction, electron microscopy) show that in this mode there is no complete annealing of defects. Apparently, the annealing mode should be chosen taking into account what has been said, for which material used alloyed by the nuclear transmutation method is used. One of the main criteria, obviously, should be considered the nominal resistivity and the uniform distribution of the introduced impurities. With insufficient annealing, the manifestation of the electrical activity of the impurity will be partially masked by the compensating effect of unannealed radiation defects. At high annealing temperatures, diffusion redistribution of the introduced impurity is possible. There are absolutely no experimental data that could answer the question of whether, upon annealing, the features associated with radiation-accelerated diffusion of the impurity are manifested (see Chapter 3).

In conclusion, we will compare the transmutation doping method with the traditional ones. What are the advantages of the transmutation doping method?

1. First of all, the predictability of doping results and ease of process control. Indeed, the concentration of the introduced impurity is determined by nuclear data (see Chap. 2) and the duration of irradiation and does not depend on the irradiation temperature or on the quality and condition of the material.

2. The uniformity of doping. When considering this issue, one should proceed from specific conditions — the type of irradiation, semiconductor material, and the necessary region of uniform doping. When irradiated by charged particles (protons, deuterons, etc.), depending on their energy, one

should expect uniform doping of only layers of units and tens of microns. Therefore, uniform doping of ingots is possible only when using neutron reactions and photonuclear reactions. The depth distribution of the introduced impurity (in the direction of the particle flow) is written as follows:

$$N(x) = \Phi_0 N_0 k\sigma \exp(-N_0 k\sigma x)$$

where Φ_0 - integral particle flow, cm^{-2}; N_0 - concentration of atoms of a substance, cm^{-3}; k - content of the required isotope, %; σ - cross section for the interaction of the reaction leading to the formation of an impurity, cm^2.

With sample thickness $d \ll (N_o k\sigma)^{-1}$ we get a uniformly doped sample. If the crystal thickness is sufficiently large, then the integrated flux Φ_0 will also be a function of depth x. In the case of silicon, in reactions involving neutrons, the uniformly doped region reaches tens of centimeters, and in semiconductor compounds including components such as indium, cadmium, the uniformly doped zone is only a fraction of a millimeter. To obtain uniformly doped silicon, it is necessary that the crystals do not contain "biographical" defects (see Chapter 4).

3. There are no fundamental restrictions on the concentration of impurities. Indeed, the restriction is only the concentration of the isotope, the transmutation of which occurs, but the concentration is percent of the total number of atoms of the substance, so that in principle an isotopically pure material can be used. The limitation is transmutation reactions with an impurity.

4. Using the screens it is possible to create the desired configuration of the doped area.

5. Silicon obtained by transmutation doping has high resistance to subsequent heat treatments and radiation.

Among the limitations of the method, we point out the following.

1. The introduction of radiation defects during irradiation. This circumstance necessitates the subsequent annealing of radiation defects. During annealing, conditions are found under which accelerated diffusion of the impurity into the effluents is possible. This process leads to an uneven distribution of impurities.

2. In the process of doping with the method of nuclear transmutations, unstable isotopes are formed and induced radioactivity arises. The background of the induced radioactivity decreases sharply when the crystals

are thoroughly cleaned of impurities giving isotopes with a long half-life. The presence of induced radioactivity leads to the fact that the doped material requires exposure to "highlight" the active isotopes before putting it into production.

3. For some materials, in particular silicon, there is a purely practical limitation - exposure time. Therefore, the method under consideration is not very suitable for introducing impurity concentrations higher than 10^{16} cm^{-3} (phosphorus in silicon).

4. The type of impurity introduced (for more details see chapter 2).

The method of transmutation doping is widely used when doping silicon with phosphorus due to (n, γ) reactions. For other semiconductor materials, in the best case, the principal possibility of doping with the nuclear transmutation method is shown.

References:

1. Rodes R. G. *Nesovershenstva i aktivnye tsentry v poluprovodnikakh* [Imperfections and active centers in semiconductors]. Moscow, Metallurgy, 1968, 371 p. (in Russian)
2. Boltaks B. I. *Diffuziia v poluprovodnikakh* [Semiconductor Diffusion]. Moscow, Fizmatgiz, 1961. 462 p. (in Russian)
3. Boltaks B. I. *Diffuziia i tochechnye defekty v poluprovodnikakh* [Diffusion and point defects in semiconductors]. St. Petersburg, Science, 1972, 384 p. (in Russian)
4. Dzhafarov T. D. *Defekty i diffuziia v epitaksial'nykh strukturakh* [Defects and diffusion in epitaxial structures]. St. Petersburg, Science, 1978, 207 p. (in Russian)
5. Fritsshe K. *Poluchenie poluprovodnikov* [Semiconductor production]. Moscow, World, 1964. 205 p. (in Russian)
6. *Atomnaia diffuziia v poluprovodnikakh* [Atomic diffusion in semiconductors]. Moscow, World, 1975. 684 p. (in Russian)
7. Fistul' V. I. *Sil'no legirovannye poluprovodniki* [Heavily Doped Semiconductors]. Moscow, Science, 1967. 415 p. (in Russian)
8. Palatiik L. S, Papirov I. I. *Epitaksial'nye plenki* [Epitaxial films], Moscow, Science, 1971. (in Russian)

9. Meier Dzh., Erikson L., Devis Dzh. *Ionnoe legirovanie poluprovodnikov* [Ion doping of semiconductors]. Moscow, Science, 1973, 296 p. (in Russian)
10. *Tekhnologiia ionnogo legirovaniia* [Ion doping technology]. Moscow, Soviet radio, 1974. 158 p. (in Russian)
11. Zorin E. I., Pavlov P. V., Tetel'baum D. I. *Ionnoe legirovanie poluprovodnikov* [Ion doping of semiconductors]. Moscow, Energy, 1975. 128 p. (in Russian)
12. *Fizicheskie protsessy v obluchennykh poluprovodnikakh* [Physical processes in irradiated semiconductors]. Novosibirsk, Science, 1977. 256 p. (in Russian)
13. Chukichev M. V., Vavilov V. S. Obrazovanie defektov reshetki pod deistviem teplovykh neitronov pri obluchenii monokristallov kremniia v iadernom reaktore [The formation of lattice defects under the influence of thermal neutrons upon irradiation of silicon single crystals in a nuclear reactor]. *Solid state physics*, 1961, vol. 3, № 5, p. 1522-1527. (in Russian)

Chapter 2. Physical fundamentals of nuclear doping of semiconductors

2.1. Nuclear reactions as a source of impurity atoms

Since the discovery of nuclear reactions, it has become a real dream for alchemists to obtain valuable materials from less valuable ones. Only unlike the methods that alchemists tried to solve this problem, is this achieved by irradiating the starting materials with particular particles: elementary (electrons, protons, neutrons, etc.) or complex (deuterons, α-particles, multiply charged ions, etc.).

If the irradiation conditions provide for the occurrence of certain nuclear reactions, then as a result of the corresponding nuclear transformations of some of the atoms of the irradiated substance, it is possible to form atoms of other chemical elements that are different from the atoms of the original substance. The interspersing of new elements can, in principle, be separated from the original substance, and the whole question comes down to whether it is worth doing because of the difficulty of obtaining and isolating the "new" substance. Nevertheless, at present, only so isotopes of transuranic elements are artificially obtained and studied, only this way on an industrial scale is obtained from irradiated nuclear fuel - uranium - other nuclear fuel - plutonium, as well as isotopes of some rare elements.

However, in most other cases, when other methods of obtaining the desired substance are known, the use of nuclear transmutations for these purposes is disadvantageous, since the concentration of the "new" substance in the irradiated material is usually very small. Therefore, nuclear transmutations accompanying most cases of natural or artificial exposure of substances that are encountered in practice are considered to be a side, most often harmful effect, which has no practical perspective, except for some exotic ones, like the cases mentioned above. Such an opinion seems truly substantiated, while we are talking about the isolation and use of the products of nuclear transmutations, regardless of the source material. But if we take into account that modern radiation technology is based on a directed change in the properties of the starting material as a result of irradiation, it is necessary to consider from this point of view the influence of the products of nuclear transmutations.

For technically and economically reasonable times of irradiation of various materials, the concentration of impurity atoms formed as a result of nuclear transmutations can be at best fractions of a percent. Therefore, the resulting atoms as a chemical impurity cannot have a significant effect on the properties of the initial material. In this regard, it is of particular interest that in some cases the practically important properties of irradiated materials, in particular, determined by the electronic structure, depend significantly on the nature and distribution of a small amount of impurity atoms. Examples of this effect are the coloration of irradiated alkali halide crystals and various glasses, as well as a change in the electrophysical properties of semiconductors.

The task of doping semiconductors is to introduce a given concentration and ensure a given distribution of a particular dopant in the bulk of the source material. This task requires the creation of optimal doping conditions for each specific semiconductor material, both by conventional methods and by the method of nuclear transmutations.

The fundamental difference between nuclear doping is that doping impurities are not introduced into the source material from the outside, but are formed during irradiation directly from the atoms of the doped material. To assess the fundamental possibility of such a process, without special experience, based on general information about the physics of nuclear radiation and their interaction with matter, one can intuitively proceed from the fact that under certain conditions each type of radiation can cause nuclear transmutations in the irradiated substance.

Depending on the type and energy of the irradiating particles, on the composition of the irradiated substance, nuclear reactions at the nuclei of various chemical elements can occur simultaneously with the formation of various impurities, so that the final doping result will be determined by the total role of all impurities. At the same time, at least part of the impurities will be the nuclei of radioactive elements and, as a result, induced radioactivity of the doped material will appear, which complicates or makes impossible the safe work of a person with this material. In addition, as a result of collisions irradiating particles with the atoms of the irradiated material, the latter transfers energy, in many cases sufficient to displace them from the nodes of the crystal lattice and even to form significant areas of damage in the lattice as a result of repeated collisions of particles with atoms and primary knocked out atoms with other atoms. The radioactive decay of unstable nuclei is an additional source of radiation disturbances (defects) created by both particles emitted from the nuclei and the decaying nuclei themselves, which can be displaced due to the recoil energy transferred to the nuclei when the particles escape. A priori, it is not clear how a complex set of radiation defects will affect the target doping result, and from general considerations about the connection of the properties of semiconductors with the perfection of their structure, it should be expected that this effect should be harmful.

Thus, already from these arguments it follows that for each specific material the center of gravity of the nuclear doping problem shifts to the range of selection of the necessary nuclear reactions and the creation of conditions for their effective without complicating the influence of other possible reactions and side effects of radiation, such as radiation defects and induced radioactivity. Let us consider the main laws governing the role of these factors and the possibility of controlling them. Thus, from these considerations, it follows that for each specific material the center of gravity of the nuclear doping problem shifts to the range of choice of the necessary nuclear reactions and the creation of conditions for their effective occurrence without the complicating effect of others possible reactions and side effects of radiation, such as radiation defects and induced radioactivity. Consider the basic laws that determine the role of these factors and the ability to manage them.

According to modern concepts (see, for example, [1, 2]), nuclear reactions leading to the formation of dopants can occur when irradiated with charged particles (protons, deuterons, α particles, ions of other elements), neutrons and γ-quantum. Schematically, the nuclear interaction of particle a with core X, as a result of which another particle b and core Y are formed, can be written as

$$a + X \rightarrow b + Y, \text{ or } X(a,b)Y \tag{2.1}$$

moreover, in all nuclear transmutations the laws of conservation of energy, momentum, total electric charge of particles and the total number of nucleons are satisfied. If we denote by Z and A, respectively, the charge and mass number of particles and nuclei participating in the reactions, then taking into account these laws

$$X_{Z_X}^{A_X} + a_{Z_a}^{A_a} \to b_{Z_b}^{A_b} + Y_{Z_Y}^{A_Y} + |Q| \qquad (2.2)$$

where Q – reaction energy, which is the difference in the energy of the set of initial and final reaction products.

It follows from (2.2) that, up to a difference in energy, the final core is obtained from the initial as follows:

$$Y_{Z_Y}^{A_Y} = X_{Z_X+Z_a-Z_b}^{A_X+A_a-A_b}$$

i.e., depending on A_a, A_b, Z_a, Z_b, the core Y can be the core of one of the isotopes of the original element X or the core of one of the isotopes of neighboring elements. In some cases, unstable isotopes are formed, which turn into stable products as a result of subsequent radioactive decay, for the description of which the same reaction scheme is applicable (2.2).

As for the reaction mechanism described by expressions (2.1) and (2.2), in many cases the energy of the particle a that has fallen into the nucleus X_{Zx}^{Ax} redistributed between the particles of the nucleus, resulting in the formation of an excited compound nucleus X_{Zx+Za}^{Ax+Aa}, which lives for some time and then decays in one of several possible ways, including the emission of gamma rays, neutrons, or other particles. The cross section of a nuclear reaction with the emission of a particle or quantum b can be represented as

$$\sigma(a,b) = \sigma_X \Gamma_b / \Gamma \qquad (2.3)$$

where σ_x - particle capture cross section a with the formation of a composite nucleus, Γ_b/Γ determines the relative probability of its decay with the emission of a particle or quantum b. Each possible decay of a compound nucleus is characterized by a partial average lifetime τ_b and the corresponding partial energy level width of the compound nucleus $\Gamma_b = \hbar/\tau_b$, full width $\Gamma = \Sigma\Gamma_b$.

The implementation of one or another variant of the decay of a compound nucleus is determined by the nature of the nuclei of the irradiated substance, as well as the nature and energy of the irradiating particles. The most important characteristic of nuclear reactions, determining the very possibility of their implementation, is the reaction energy

$$Q = E_{R1} - E_{R2} = E_{K2} - E_{K1} \qquad (2.4)$$

where E_R и E_K - respectively, rest energy and kinetic energy, and indices 1 and 2 relate to the initial and final states of the nuclear reaction

$$E_{R1} = M_X c^2 + M_a c^2; E_{R2} = M_Y c^2 + M_b c^2$$

$$E_{K1} = E_{kX} + E_{ka}; E_{K2} = E_{kY} + E_{kb}$$

If $Q>0$, then the nuclear reaction can occur at any kinetic energy of the irradiating particles, and in the case of $Q<0$, the value of E_K must exceed the threshold value

$$(E_k)_{\min} = \frac{M_X + M_a}{M_X}|Q| \tag{2.5}$$

moreover $(E_K)_{min}$ exceeds $|Q|$ by the value of the kinetic energy of the composite core.

Exoenergy ($Q>0$) are nuclear reactions on slow neutrons and some others. And, on the contrary, many reactions on charged particles are endoenergetic ($Q<0$). In particular, to illustrate the threshold character of (α, n) reactions, Table 2.1 shows the $(E_K)_{min}$ and Q values for silicon isotopes [3].

Table 2.1. Energy characteristics of nuclear reactions on silicon isotopes [3]

Isotope	Nuclear reactions and conversion products	Q, MeV	$(E_K)_{min}$, MeV
Si^{28}	$Si_{14}^{28}(\alpha,n)\,S_{16}^{31^{\beta+}} \rightarrow P_{15}^{31}$	-7.72	8.82
Si^{29}	$Si_{14}^{29}(\alpha,n)\,S_{16}^{32}$	-1.54	1.75
Si^{30}	$Si_{14}^{30}(\alpha,n)\,S_{16}^{33}$	-3.49	3.95

Of these, as well as those presented in table 2.2-2.5 of the data it follows that if for the introduction of a specific impurity it is necessary to use certain particles, the corresponding radiation source should provide for the selected reaction to obtain a sufficiently intense beam of such particles with an energy above the threshold. It is desirable that all particles in the beam have the same energy, since in this case it is possible to reliably carry out a theoretical calculation and then practically implement the optimal irradiation conditions.

2.2. Nuclear reactions by charged particles

Under the action of charged particles of sufficiently high energy, nuclear reactions of the types (α, p), (α, n), (p, α), (p, n), (p, γ), (p, d) and other. The threshold nature of these reactions is determined by the fact that the energy of a charged particle incident on the nucleus must be sufficient to overcome the mutual Coulomb repulsion of the particle and the nucleus, and in the event of the subsequent emission of another charged particle from the excited composite nucleus, it must also receive sufficient energy in the composite nucleus to overcome this Coulomb barrier. If r_n is the radius of the nucleus, and e is the electron charge, then the height of the barrier is determined by the expression

$$B_k = \frac{Z_X Z_n e^2}{r_n} \approx \frac{Z_X Z_a}{A_X^{1/3}} \text{ MeV} \tag{2.6}$$

and reaches 5-10 MeV for light nuclei, 10-20 MeV for medium nuclei and 20-30 MeV for heavy nuclei [2].

It should be noted that, according to the laws of quantum mechanics, there is a probability of overcoming the Coulomb barrier even when the particle energy is lower than E_k, but this is not of fundamental importance to describe the essence of threshold phenomena.

Historically, the first sources of charged particles, namely α-particles, were radioactive elements Ra, Rn, Po, Pu and others, which as a result of radioactive decay emit in 1 s per 1 g of source up to 10^{10}-10^{11} α-particles [4] with energies in the range of 4-8 MeV [3-5].

In addition to non-monochromaticity, the energy and intensity of the α-particle beams are too low for real semiconductor materials to be doped. Since only a small fraction of charged particles ($\leq 10^{-3}$ [2]) induces nuclear reactions passing through the irradiated material, to solve practical problems of nuclear doping, charged particle beams are required whose energy exceeds threshold values for the nuclei of doped elements, and the intensity should be several orders of magnitude higher than the radiation intensity of the above isotopes. This is necessary in order to introduce an integral particle flux higher than 10^{17}-10^{19} cm^{-2} into a doping zone in a reasonable time of exposure and to obtain in the unit volume of the material the amount of nuclear transmutations comparable with the concentration of the desired impurity. Currently, there are technical possibilities for solving this problem with the help of linear and cyclic accelerators of charged particles, which make it possible to obtain monoenergetic beams of electrons, protons, deuterons and α particles.

Without going into further discussion of the issue, we assume that there is a beam of charged particles suitable for a particular doping problem. The question then becomes how this beam transfers its energy to the atoms of the irradiated material. It turns out that due to the Coulomb interaction with the medium, the kinetic energy of charged particles is mainly spent on the ionization and excitation of atoms of the irradiated substance (ionization drag). Estimates show (see [2]) that the cross section for processes determined by ionization (~ 10^{-21} cm^2) is approximately 1000 times larger than the cross section for nuclear interaction (~ 10^{-24} cm^2). Therefore, along the entire path (R_0) of the particles, when they decelerate to zero energy, their flow can be considered almost constant. Strictly speaking, for a monoenergetic particle beam, R_0 has the mean free path, with respect to which there is a scatter of R values described by a Gaussian distribution.

On the other hand, as the substance penetrates deep into the substance, the particle energy (E) decreases by about 3.5 eV during each ionization event [2], and the probability of nuclear interaction depends on the energy that the particle has at the moment of interaction. Therefore, the number of nuclear interactions in a thin layer dx at a depth x from the target surface can be represented as

$$dv = \eta(x)N\sigma(x)dx \approx \eta_0 N\sigma(x)dx_y \tag{2.7}$$

where N - concentration of target nuclei; η_0 and η - particle flow at the surface and at a depth x, respectively; $\sigma(x)$ - interaction cross section. Hence, the total number of interactions in a layer of matter whose thickness is greater than R_0 is

$$v = \eta_0 N\int_0^{R_0} \sigma(x)dx = \eta_0 N\int_0^{R_0} \frac{\sigma(E)dE}{\left|\frac{dE}{dx}\right|} \tag{2.8}$$

where E_0 - initial particle energy.

Therefore, the yield of a nuclear reaction, determined by the fraction of particles that have experienced nuclear interaction,

$$V(E) = \frac{v}{\eta_0} = N\int_0^{R_0} \frac{\sigma(E)dE}{\left|\frac{dE}{dx}\right|} \tag{2.9}$$

i.e., the yield of a nuclear reaction at a charged particle energy equal to E, defined by section $\sigma(E)$ and specific ionization of the particle dE/dx. And conversely, knowing the function dE/dx and dV/dE, from (2.9) can determine the cross section of the interaction

$$\sigma(E) = \frac{1}{N}\frac{dV}{dE}\left|\frac{dE}{dx}\right| \tag{2.10}$$

It is known (see [2, 6]) that for a charged particle the quantity dE/dx proportional to the squared particle charge, electron concentration in the medium (n_e), some function of speed $f(v)\sim 1/v^2$ and

$$\frac{dE}{dx} \sim Z^2 n_e f(v) \tag{2.11}$$

does not depend on the mass of the particle, i.e., the *dE/dx* dependence allows one to recalculate the data on the motion of one particular particle in a particular medium on the motion of other particles in a specified or in other media.

This is achieved by comparing the same values *dE/dx* when sequentially changing individual parameters included in (2.11), while maintaining other parameters unchanged. For example, when a proton (Z = 1) and an α-particle (Z = 2) move in the same medium ($n_e = const$)

$$\left(\frac{dE_a}{dx}\right)_{E=E_0} = 4\left(\frac{dE_p}{dx}\right)_{E=E_0/4} \tag{2.12}$$

Since for a specific particle and a certain medium, *dE/dx* depends only on the velocity and, therefore, on the particle energy, i.e., *dE/dx=f(E)*, integrating this expression over all energy values, we can obtain the total particle path

$$R = \int_0^{E_0} \frac{dE}{f(E)} \tag{2.13}$$

For practical calculations, you can use the dependence of the specific energy loss of various particles in air on the energy of these particles. With the help of such dependencies given in the form of graphs in books [5, 6], it is possible to determine the mean free path of particles first in air and then in other media. For example, the range of α-particles (R_α, x) in the material X_Z^A can be determined by the range in air (R_α, air) using the empirical formula [1, 6]

$$R_{a,X}(E) = 0,56 R_{\alpha,air}(E) A^{1/3} \tag{2.14}$$

where is the value $R_{\alpha,air}$ in centimeters refers to air at temperature 15°C and pressure 760 mm Hg Art, and $R_\alpha x$ obtained in units of mg/cm^2.

The range of protons is related to the range of α particles by another formula valid for $E \geq 0,5$ MeV:

$$R_p(E) = 1,007 R_\alpha(3,972)E - 0,2 \tag{2.15}$$

where $R_\alpha(3,972)$ - range of α particles at energy 3,972. The range of other singly charged particles with mass M_x (except for the electron) is related to the range of protons by the ratio:

$$R_x(E) = \frac{M_x}{M_p} R_p \left(\frac{M_p}{M_x} E \right) \quad (2.16)$$

Due to the rapid deceleration of charged particles, they can be used to obtain doped layers of small thickness with a very uneven distribution of dopants across the thickness. Nevertheless, the fundamental possibility of doping germanium with the help of α particles and deuterons [7] and silicon with the help of α particles with an energy of 27.2 MeV [8] was described. Possible nuclear reactions of this type and the resulting dopants are given in Tables 2.2 and 2.3, which also reflect the nature of the radioactivity and half-life of unstable isotopes. Under the action of these reactions, *p*-type silicon with a resistivity of ρ~30 Ohm.cm at a depth of 240-340 μm a layer with an *n*-type conductivity appeared [8]. Similarly, impurities Ga, Se, As, and Br can be introduced into germanium. However, there is not enough data on the effectiveness of the reactions given in Tables 2.2, 2.3, so there is no experience in the world practice of developing and applying the method of doping semiconductor materials using charged particles. It is possible that the progress of semiconductor electronics may in the near future make it expedient to selectively dope thin layers.

In particular, with the help of charged particles, it is possible to obtain flat doped layers of any configuration using absorbing screens of an appropriate shape. In this case, a sharp boundary can be obtained between the doped and undoped portions of the semiconductor, since thin layers of generally accessible protective materials can completely inhibit the flow of charged particles.

Table 2.2. Nuclear reactions in silicon by α-particles

Characterization of nuclear reactions and dopants	Threshold energy, MeV
$Si_{14}^{28}(\alpha,\gamma)S_{16}^{32}$	-
$Si^{29}(\alpha,\gamma)S^{33}$	-
$Si^{30}(\alpha,\gamma)S^{34}$	-
$Si^{29}(\alpha,n)S^{32}$	1,75
$Si_{14}^{28}(\alpha,p)P_{15}^{31}$	2,08
$Si^{29}(\alpha,p)P^{32}\xrightarrow[14,3\text{ days}]{\beta^-}S^{32}$	2,78
$Si^{30}(\alpha,p)P^{33}\xrightarrow[24,4\text{ days}]{\beta^-}S^{33}$	3,35

$Si^{30}(\alpha,n)S^{33}$	3,95
$Si^{28}(\alpha,n)Si^{31}\xrightarrow[2,7\ sec]{\beta^{+}}P^{31}$	8,82
$Si^{30}(\alpha,2p)Si^{32}\xrightarrow[650\ years]{\beta^{-}}P^{32}\xrightarrow[14,3\ days]{\beta^{-}}S^{32}$	11,99
$Si^{29}(\alpha,2p)Si^{31}\xrightarrow[2,6\ hours]{\beta^{-}}P^{31}$	12,62
$Si^{29}(\alpha,np)P^{31}$	12,79
$Si^{30}(\alpha,2n)S^{32}$	14,80
$Si^{28}(\alpha,np)P^{32}\xrightarrow[14,3\ days]{\beta^{-}}S^{32}$	14,81
$Si^{29}(\alpha,2n)Si^{32}\xrightarrow[2,7\ sec]{\beta^{+}}P^{31}$	20,3
$Si^{30}(\alpha,2pn)Si^{31}\xrightarrow[2,6\ hours]{\beta^{-}}P^{31}$	24,6
$Si^{30}(\alpha,2np)P^{31}$	25,5

Note: Nuclear reactions under the influence of α-particles lead to the appearance of donors in silicon.

Table 2.3. Nuclear reactions in germanium under the action of charged particles [5, 7]

Reactions and nuclear reaction products	Nature of the effect of impurities
$Ge_{32}^{70}(d,p)Ge_{32}^{71}\xrightarrow[11,4\ days]{K-capture}Ge_{31}^{71}$	a
$Ge^{72}(d,p)Ge^{73}$	n
$Ge^{73}(d,p)Ge^{74}$	n
$Ge_{32}^{74}(d,p)Ge_{32}^{75}\xrightarrow[82min]{\beta^{-}}As_{33}^{75}$	d
$Ge_{32}^{76}(d,p)Ge_{33}^{77}\xrightarrow[11,3\ hours]{\beta^{-}}As_{33}^{77}\xrightarrow[38,7\ hours]{\beta^{-}}Se_{34}^{77}$	d
$Ge^{70}(d,n)As^{71}\xrightarrow[62hours]{K-capture}Ge^{71}\xrightarrow[11,4days]{K-capture}Ga^{71}$	a
$Ge^{72}(d,n)As^{73}\xrightarrow[80,3\ days]{K-capture}Ge^{73}$	n
$Ge^{73}(d,n)As^{74}\xrightarrow[17,9days]{\beta^{+}}Ge^{74}$	n
$Ge^{74}(d,n)As^{75}$	d
$Ge^{29}(d,n)As^{77}\xrightarrow[38,7\ hours]{\beta^{-}}Se^{77}$	d
$Ge^{70}(\alpha,n)Se^{73}\xrightarrow[7,1hours]{\beta^{+}}As^{73}\xrightarrow[80,3days]{K-capture}Ge^{73}$	n
$Ge^{72}(\alpha,n)Se^{75}\xrightarrow[120,4\ days]{K-capture}As^{75}$	d
$Ge^{73}(\alpha,n)Se^{76}$	d
$Ge^{74}(\alpha,n)Se^{77}$	d
$Ge_{32}^{76}(\alpha,n)Se_{34}^{79}\xrightarrow[6,5\cdot 10^{4}\ years]{\beta^{-}}Br_{35}^{79}$	d
$Ge^{70}(\alpha,2n)Se^{72}\xrightarrow[8,4\ days]{K-capture}As^{72}\xrightarrow[26hours]{\beta^{+}}Ge^{72}$	n

$Ge^{72}(\alpha,2n)Se^{74}$	d
$Ge^{73}(\alpha,2n)Se^{73}\xrightarrow[120,4\ days]{K-capture}As^{73}$	d
$Ge^{74}(\alpha,2n)Se^{76}$	d
$Ge^{74}(\alpha,2n)Se^{75}$	d
$Ge^{70}(\alpha,p)As^{73}\xrightarrow[80,3\ days]{K-capture}Ge^{73}$	n
$Ge^{72}(\alpha,p)As^{75}$	d
$Ge^{73}(\alpha,p)As^{76}\xrightarrow[26,4 hours]{\beta^-}Se^{74}$	d
$Ge^{74}(\alpha,p)As^{77}\xrightarrow[38,7 hours]{\beta^-}Se^{77}$	d
$Ge^{76}(\alpha,p)As^{79}\xrightarrow[9 min]{\beta^-}Se^{79}\xrightarrow[6,5\cdot 10^4\ years]{\beta^-}Br^{79}$	d

Note. Here and in subsequent similar tables, a is an acceptor; n is a neutral atom; d — donor.

2.3. Nuclear reactions under the influence of γ-rays

Threshold reactions include a group of nuclear fission reactions by γ rays (photonuclear reactions), namely, (γ, *n*) - (γ, *p*) - and (γ, *α*) reactions. These reactions are always endoenergetic, and for the implementation of such reactions, it is necessary that the γ-quantum energy exceeds the separation energy of the corresponding particles. In the case of (γ, *n*) - reactions, the threshold energy practically coincides with the binding energy of the neutron in the nucleus, and in the case of charged particles being pulled out of the nucleus, it is necessary to give them additional energy to overcome the Coulomb barrier. Therefore, in the general case, the probability of (γ, *n*) reactions is higher than for (γ, *p*) and (γ, α) reactions, and in the region of relatively low γ-ray energies (~ 10 MeV) of the reaction (γ, *n*) prevail. At a γ-ray energy of ~ 100 MeV, reactions with the release of several particles are possible, that is, reactions of the type (γ, 2*n*), (γ, *pn*), etc. The ratio of the outputs (γ, *p*) - and (γ, *n*) - reactions obtained in the experiment, V(γ,*p*)/V (γ, *n*)=10^{-2} [2], whereas, according to the ideas about the occurrence of photonuclear reactions with the formation of an intermediate nucleus, this ratio would have a value of 10^{-5}-10^{-4}.

This feature of photonuclear reactions also manifests itself during the reaction cross section. In the range of γ-ray energies of 10–20 MeV, the cross section has a broad (Γ = 3–7 MeV) resonance maximum (giant resonance), the position of which for different nuclei changes according to the law $(E_\gamma)_{res}$

~ $A^{-1/6}$ and which is explained by dipole vibrations nuclei under the action of gamma radiation.

Since photonuclear reactions can create impurities in almost any material, they are of interest also for doping semiconductors. This interest is mainly determined by the fact that, in contrast to charged particles, γ-quantum have high penetrating power, and, therefore, with their help it is possible to ensure uniform doping of large volumes of material with a certain set of impurity atoms. Table 2.4 shows the data (see [4]) on neutron binding energies in the nuclei of elements that are components of a number of semiconductor materials.

Table 2.4 Neutron binding energy in some nuclei

	A	E_n, MeV
Si_{14}	28	17,1780±0,14
	29	8,4777±0,0067
	30	10,6142±0,0078
	31	6,5921±0,00909
	32	310±0,055
P_{15}	29	17,32±0,31
	30	11,327±0,021
	31	12,316±0,012
	32	7,9367±0,0041
S_{16}	32	15,081±0,018
	33	8,6425±0,004
	34	11,4208±0,006
	35	6,9818±0,0059
	36	9,881±0,0120
	37	4,398±0,099
	38	7,89±0,24
Ga_{31}	65	11,75±0,17
	67	9,127±0,047
	67	11,217±0,042
	68	8,279±0,021
	69	10,228±0,036
	70	7,730±0,042
	71	9,194±0,063
	72	6,968±0,095
	73	9,01±0,13
	74	6,02±0,29
In_{49}	109	10,34±0,26
	110	7,90±0,22
	111	9,69±0,30
	112	7,92±0,2
	113	9,34±0,19
	114	7,31±0,18
	115	9,03±0,19
	116	6,60±0,27
	117	9,00±0,34

Element	A	E_n, MeV
Sb_{51}	117	10,06±0,53
	120	7,09±0,32
	121	9,29±0,26
	122	6,827±0,025
	123	8,956±0,025
	124	6,45±0,25
	125	8,68±0,24
Ge_{32}	68	12,01±0,7
	69	8,657±0,63
	70	11,617±0,047
	71	7,311±0,066
	72	11,19±0,095
	73	6,57±0,12
	74	10,12±0,12
	75	6,50±0,010
	76	9,45±0,14
	77	5,96±0,16
	78	8,92±0,22
As_{33}	70	8,97±0,40
	71	11,84±0,15
	72	8,84±0,11
	73	10,56±0,13
	74	7,93±0,12
	75	10,230±098
	76	7,290±0,093
	77	9,700±0,090
	78	7,07±0,16
	79	8,77±0,21
	80	6,24±0,30
Cd_{48}	107	7,54±0,46
	108	10,42±0,22
	109	7,26±0,22
	110	9,84±0,22
	111	6,97±0,28
	112	9,29±0,27
	113	6,42±0,19

	114	9,04±0,18
	115	6,16±0,19
	116	8,64±0,40
	117	5,87±0,65

The prospects of this direction are now also supported by the possibility of obtaining γ-quantum with any energy in the form of bremsstrahlung of electrons. For this, monoenergetic electrons with energies in the range of 25-60 MeV, obtained with the help of accelerators, are sent to a target of heavy metals (Pb, Bi, W, U, etc.). As a result of electron deceleration, a continuous spectrum of γ-radiation is formed, the maximum energy of which is equal to the kinetic energy of electrons E_k, and the intensity of γ-radiation is approximately inversely proportional to the energy of γ-quantum (Fig. 2.1).

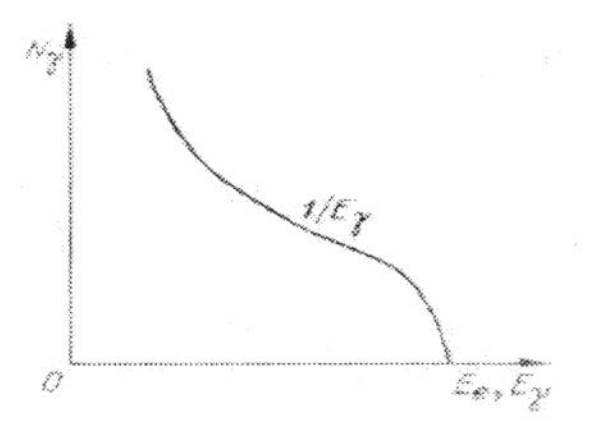

Figure 2.1. The spectrum of gamma radiation arising from the deceleration of electrons with an energy E_e on an accelerator target

Despite the fact that there are difficulties associated with the interpretation of the integral effect of γ-quantum of different energies, as well as with the need to protect against this radiation, in recent years, the promise of using bremsstrahlung γ-radiation of electron accelerators to simulate the radiation effect of neutrons and protons on silicon [9], and for direct doping of silicon [10].

Possible nuclear reactions (γ, *n*), (γ, *p*) and (γ, *α*) on silicon isotopes, which can lead to the formation of acceptor impurities Al^{27} and magnesium isotopes, with Al^{27} corresponding to a level of 0.057 eV from the ceiling of the valence band, and Mg mixtures give deep levels in the band gap [10], are presented in table 2.5. Unfortunately, there are no data on threshold energies (except for (γ, *n*)-reactions) and cross sections of all reactions listed in Table 2.5 at different γ-quantum energies. If such data were available, it would be possible to determine the concentration of Al^{27} (and similarly for other impurities) according to the formula [10]

$$N_{Al} = k_i N_{Si} t \left[\int_{E_{thr(\gamma,n)}}^{E_{\max}} \sigma_n(E)\Phi(E)dE + \int_{E_{thr(\gamma,p)}}^{E_{\max}} \sigma_p(E)\Phi(E)dE \right]$$

where k_i - relative concentration of the isotope Si^{28} (92,2%); N_{Si} - silicon atom concentration; *t* - exposure time; $E_{thr(\gamma,n)}$,$E_{thr(\gamma,p)}$, $\sigma_n(E)$ и $\sigma_p(E)$ - threshold energies and differential cross sections for reactions (γ,n) and (γ,p), respectively; Φ(E) - differential radiation flux, E_{max} - maximum energy of γ radiation.

A similar analysis can be done for all semiconductor materials, and in each case, choose the energy of γ-quantum, which provide the most effective course of those photonuclear reactions, which will lead to the predominant formation of useful dopants.

Table 2.5. Characteristics and products of some photonuclear reactions [5, 10]

Photonuclear reactions and conversion products	The nature of the influence of impurities
$Si_{14}^{28}(\gamma,n)Si_{14}^{27}\xrightarrow[4,1\,sec]{\beta^+}Al_{13}^{27}$	a
$Si^{29}(\gamma,n)Si^{28}$	n
$Si^{30}(\gamma,n)Si^{29}$	n
$Si^{28}(\gamma,p)Al^{27}$	a
$Si^{29}(\gamma,p)Al^{28}\xrightarrow[2,31\,min]{\beta^-}Si^{28}$	n
$Si^{30}(\gamma,p)Al^{29}\xrightarrow[6,6\,min]{\beta^-}Si^{29}$	n
$Si_{14}^{28}(\gamma,\alpha)Mg_{12}^{24}$	a
$Si^{29}(\gamma,\alpha)Mg^{25}$	a
$Si^{30}(\gamma,\alpha)Mg^{26}$	a

2.4 Nuclear reactions by neutrons

The main hopes in the field of nuclear doping are currently assigned to neutrons, since they, being uncharged particles, have great penetrating power and can interact at almost all energies with almost all nuclei. In addition, various neutron sources of different intensities have been mastered and are available. One of the types of neutron sources is nuclear reactions under the influence of charged particles - (α,n), (d,n), (p,n) - reactions. In particular, the (α, n) reactions mentioned in § 2.2 in beryllium, i.e., $Be^9(\alpha,n)C^{12}$, under the action of α-particles of the radioactive isotopes Ra, Rn, Po, Pu, etc. are the basis of Ra - Be, Po - Be, Pn - Be and other similar sources that make it possible to obtain ~ 10^7 neutrons /sec per 1 g of Ra or equivalent in activity of another isotope, which corresponds to ~ 10^{-5} neutrons per α-particle [4] and confirms the conclusion made in § 2.2 about the small yield of nuclear reactions on charged particles. It should also be noted that sources of this type produce neutrons of various energies with a complex spectrum similar to that shown in Fig. 2.2 for Ra - Be and Po - Be sources [3]. With the development of accelerators of charged particles, in particular cyclotrons, the (*d,n*) reaction has become very important, by which, when irradiated with accelerated deuterons, deuterium [$D^2(d,n)He^3$], tritium [$H^3(d,n)He^4$], beryllium [$Be^9(d,n)B^{10}$], lithium [$Li^7(d,n)2He^4$] and [$Li^7(d,n)Be^8$] and other isotopes from the nuclei of the target neutrons are released.

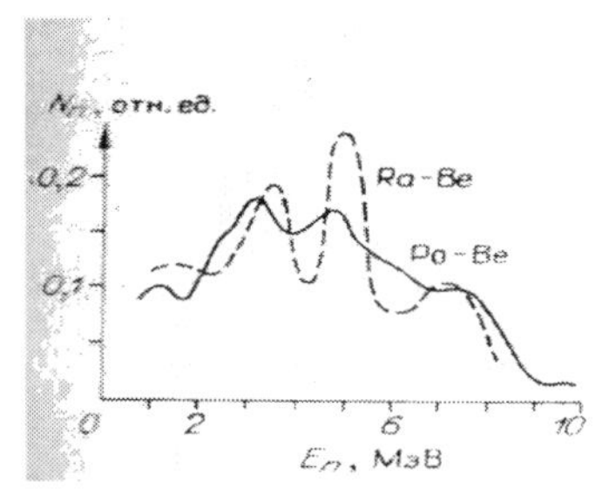

Figure 2.2. Energy spectra of neutrons emitted by Ra – Be and Po - Be sources

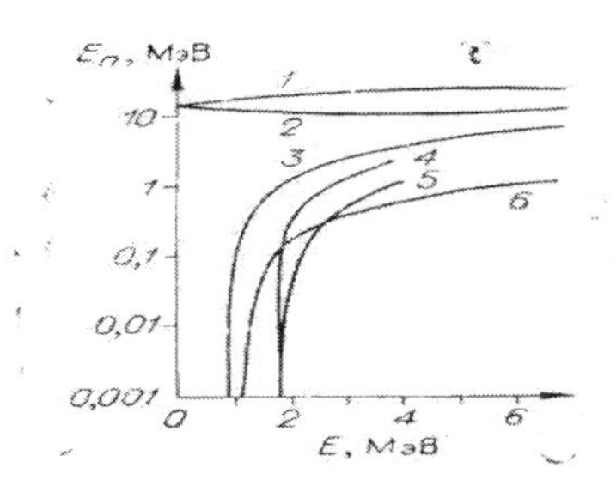
Figure 2.3. Dependence of the neutron energy En emitted at an angle θ on the energy of the bombarding particles E in various nuclear reactions. 1,2 - H^3(d,n) He^4; 3,4 - Li^7 (p,n) Be^7; 5,6- H^3(p,n) He^3;(1,3,5- θ = 0:2, 4, 6-θ = 180°

Almost all of these reactions are not threshold and can occur at low deuteron energies. In the case of reactions $D^2(d,n)He^3$ and $H^3(d,n)He^4$, the neutron energy depends on the angle of departure from the target, and in each specific direction monoenergetic neutrons fly out, the output of which corresponds to 1 neutron for every 200 deuterons with energy of 16 MeV from the Be target, $5*10^8$ neutrons per 1 μC deuteron with an energy of 600 keV from a tritium gas target (i.e., ~ $1:10^4$) and 10^8 neutrons per 1 μC deuteron with an energy of 200 keV from a Li target (see [4]). In connection with the possibility of producing monoenergetic neutrons, (*p,n*) reactions are widely used. For their implementation, monoenergetic proton beams are obtained (most often with the help of accelerators such as the Van de Graaff generator), which bombard thin lithium targets [$Li^7(p,n)Be^7$], tritium

[$H^3(p,n)He^3$], Be^{10}, C^{14}, Sc^{45}, V^{51} and other materials.

As a result of the first two reactions, one can obtain monoenergetic neutrons in a certain range of proton energies E_p, whose energy $E_n \sim E_p$ is the threshold energy equal to 1.882 and 1.019 MeV for Li^7 and tritium, respectively. The dependence of the energy of neutrons generated as a result of (*d,n*) - and (*p,n*) reactions on the energy of bombarding deuterons and protons is shown in Fig. 2.3 [5, 11]. In the region of proton energies shown in these graphs, the neutron yield in (*p,n*) reactions is of the same order of magnitude as for other charged particles, i.e., $\sim 10^{-5}$. It follows that neutron sources based on the considered (*d,n*) - and (*p,n*) - reactions allow one to obtain relatively identical maximum fluxes of the order of 10^{11} $cm^{-2}*sec^{-1}$, as in the case of the reaction $H^3(d,n)He^4$ [12]. These intensities are insufficient for the accumulation of a noticeable concentration of the products of nuclear transmutations. In addition, using these sources, neutrons with energies of 10^4–10^7 eV are obtained, which, as will be shown below, are inconvenient for nuclear doping.

The problem of obtaining intense neutron beams with a flux density of interest for the practice of nuclear doping can be solved in two ways. First, the photonuclear reactions considered in the previous section, which occur upon irradiation of heavy targets (Bi, W, U, etc.) with electrons or γ-quantum with an energy of 20-40 MeV, provide a neutron yield of $\sim 10^{14}$ $cm^{-2}\cdot sec^{-1}$ [12].

A further increase in the neutron yield is achieved by irradiating heavy targets with very fast protons with an energy of $\sim$ 1 GeV, as a result of which neutron beams of record intensity can be obtained with an average density of $\sim 10^{16}$ $cm^{-2}\cdot sec^{-1}$ (see [4]) exceeding the maximum achievable ($\sim 10^{15}$ $cm^{-2}\cdot sec^{-1}$) in the most advanced nuclear reactors. Such sources either operate or are created in industrialized countries. However, at present they are too unique and inaccessible in comparison even with nuclear reactors, although they are beginning to be used on an equal footing, and in some cases, instead of reactors in physical experiments and radiation materials science. Therefore, at present and in the near future, the main base for nuclear doping should be considered nuclear reactors, in which fast neutrons generated as a result of fission of the nuclei of a nuclear fuel, after slowing down and establishing thermal equilibrium with this medium, have an energy spectrum described by the Maxwell distribution.

A qualitatively similar procedure for the formation of the neutron spectrum is also necessary for sources based on charged particle accelerators.

A typical neutron spectrum of the WWR-t research water-water reactor is given in [12] (see Chap. 4). The reactor spectrum contains neutrons with energies from very slow ($E_n<0.01$ eV) to very fast ($E_n>10^7$ eV). Above, we considered nuclear reactions as a source of neutrons of one or another energy, and now, in connection with the doping problem, it is necessary to analyze the features of nuclear reactions that can occur in substances under the influence of these neutrons of different energies.

In the case of neutron irradiation, the cross section for the formation of the composite nucleus σ_x (see § 2.1) substantially depends on how close the energy of the incident neutron is to the eigenvalues of the energy of the levels of the composite nucleus.

In the subsequent decay of a composite nucleus in the range of neutron energies accessible with a nuclear reactor up to 10^6-10^7 eV, the wavelength of which is more than 10^{-12} cm, i.e., much larger than the radius of the nuclei, only the processes of emission of neutrons and γ-quantum are of significant importance (i.e., $\Gamma=\Gamma_n+\Gamma_\gamma$), and in this case, the most probable is the emission of a neutron with an energy equal to the energy of the original neutron ((*n,n'*-reaction or elastic neutron scattering).

In the general case, an excited composite nucleus can emit a neutron with an energy lower than the initial (*n,n'*) - reaction, or inelastic neutron scattering), and then go into the ground state with the emission of one or more γ-quantum. In the form of gamma quantum, the entire excitation energy of a compound nucleus ((*n*,γ) reaction or radiation capture of neutrons) equal to

$$E^* = \frac{M_x}{M_x + M_n} E_n + B_n \tag{2.17}$$

where M_x and M_n - core mass X_Z^A and neutron mass; E_n - kinetic energy of a neutron; B_n - binding energy of the last neutron in a compound nucleus X_Z^{A+1}.

It should be noted that the (*n,n*) - and (*n,n'*) reactions with the emission of one neutron are formally equivalent to scattering of the primary neutron without changing the structure of the target nucleus. Therefore, nuclear reactions such as a source of impurity atoms can be ignored, although they remain a source of radiation defects. In addition, the very possibility of inelastic scattering of neutrons requires that their energy $(M_x + M_n)/M_x$ times exceed the energy of the first excited Si^{28} isotope is 1.772 MeV [5]. The threshold energies of endoenergetic nuclear reactions $(n,2n)$ are even higher, amounting to 17.4 and 8.7 MeV for isotopes Si^{28} и Si^{29}, respectively [5].

Since the fraction of neutrons with energies above 2–3 MeV in the WWR type reactors in the neutron spectrum of a nuclear reactor is $\sim 10^{-2}$, the main attention should be paid to exoenergy (n,γ) reactions, which are possible at any neutron energies in practically all nuclei. In this case, the radiative capture cross section in the energy region En>>10^5 eV (fast neutrons) $\sigma\ (n,\ \gamma) \approx r_n^2 \Gamma\gamma/\Gamma$, where r_n is the radius of the nucleus. In the region of neutron energies close to the eigenvalues of the energy of the levels of the compound nucleus (resonance neutrons with $E_n \sim 0.1$-50 eV), the cross section $\sigma\ (n,\gamma)$ is described by the Breit - Wigner resonance formula and can significantly exceed the geometric cross section of the nucleus.

If the neutron energy is small compared with the energy of all resonant levels, then this is precisely the situation for neutrons with $E_n \sim 10^{-2}$-10^4 эВ [12], then section $\sigma(n,\gamma)$ usually obeys the law $1/v$, i.e.

$$\sigma(n,\gamma) = \sigma_0 \left(E_0 / E_n\right)^{1/2} = \sigma_0 \left(v_0 / v\right) \tag{2.18}$$

where σ_0, v_0 и E_0 - some constant values of the reaction cross section, velocity and neutron energy, taken as the reference point. The range of thermal energies of neutrons is of the greatest practical interest for doping, since this range accounts for almost the entire number of neutrons in the spectrum of a nuclear reactor described by the Maxwell velocity distribution

$$n(v) = \frac{4}{\sqrt{\pi}} \left(\frac{M_n}{2kT}\right)^{3/2} v^2 \exp\left(-M_n v^2 / 2kT\right) \tag{2.19}$$

where k – Boltzmann constant; T – temperature.

Averaging the cross section over the spectrum of n(v) gives

$$\bar{\sigma} = \int \frac{\sigma_0 v_0}{v} n(v) v dv \Big/ \int n(v) v dv = \frac{\sigma_0 v_0}{\bar{v}} \tag{2.20}$$

where $\bar{v} = \sqrt{8kT/\pi Mn}$ - average speed.

In the practice of calculations and measurements, it is more convenient to deal with the values of the cross section at the most probable neutron velocity

$$v_{n.b} = \sqrt{\frac{2kT}{M_n}} = \bar{v}\frac{\sqrt{\pi}}{2} = \bar{v} / 1{,}128$$

corresponding to the maximum of the distribution n(v). Then we can accept that $v_0 \equiv v_{n.b.} \sigma_0 \equiv \sigma_{n.b.}$, and take into account that at room temperature (T = 293K) $v_{n.b.}$= 2200 m/s. As a result, a relationship is achieved between the average cross-sectional spectrum of (n,γ) -reactions on thermal neutrons with the cross-sectional value, usually measured on neutrons with the most probable speed:

$$\sigma = \frac{\sigma_{n.b} v_{n.b}}{\bar{v}} = \frac{\sigma_{n.b}}{1{,}128} = \frac{\sigma_{2200}}{1{,}128} \quad (2.21)$$

The energy dependence of the cross section for silicon can be described by the following approximate equation (see [13]):

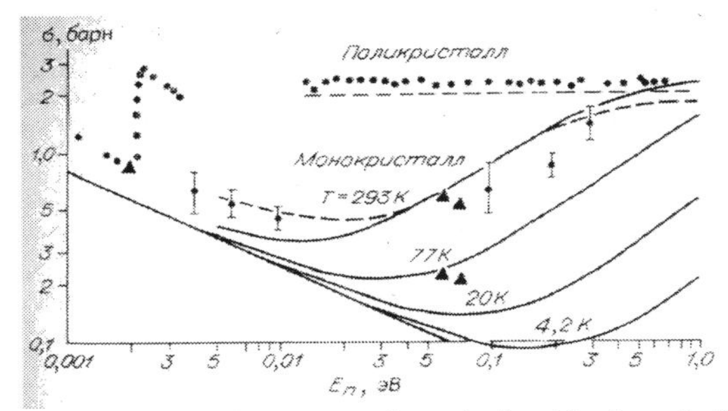

Figure 2.4. Dependence of the cross section calculated by formula (2.22) for thermal neutrons in polycrystalline and single crystal silicon at various temperatures. Dashed lines — experimental data obtained by the time-of-flight method at the WWR-t reactor.
The experimental data of the authors are also presented [13].

$$\sigma = \sigma_n + \sigma_p = \sigma_n + \sigma_n\left(1 - e^{-2w}\right) = 0{,}8/\sqrt{E_n} + 2{,}25\left(1 - e^{-CEnT}\right) \quad (2.22)$$

where E_n - neutron energy in MeV; σ_n - absorption cross section equal to 160 mbar at E_n = 25 MeV; σ_p - scattering cross section; σ_a - 2.25 barn free atom cross section; e^{-2w} - Debye – Waller factor; C = 1,439.10^{-5} - dependency constant σ (E_n) to σ = 0,55 barn with E_n = 50 MeV and T = 300K.

The energy dependences of the silicon cross section calculated from (2.22) based on [13] at different temperatures, as well as some experimental data confirming the analytical validity, dependences (2.22) are shown in Fig. 2.4. As we see, for polycrystalline silicon, the cross section in the thermal region is independent of energy, and for single crystals it varies according to the 1 / v law and depends on temperature. This must be borne in mind when strictly assessing the factors affecting the results of nuclear doping. In particular, the temperature dependence of the cross section makes it possible to reduce the role of certain reactions if doping is carried out at different temperatures.

2.5. Dopants and the nature of their distribution in semiconductors doped with (n, γ) reactions

According to the general scheme of nuclear reactions [(see (2.1)], the primary products of (n, γ) reactions are $X_Z^{(A+1)}$ nuclei one atomic unit heavier than the original. These can be heavier isotopes of the same or neighboring if these isotopes are present in a natural mixture of isotopes of the irradiated material and are stable, then the corresponding nuclear transmutations only lead to a change in the initial concentration of stable isotopes in the irradiated material, without creating impurities of neighboring elements of the periodic system.

If the primary products of (n, γ) reactions are unstable isotopes, then they undergo (in one or several stages) subsequent radioactive decay, most often β - decay, according to the scheme

$$X_Z^A(n,\gamma)*X_Z^{A+1}\frac{\beta}{T_{1/2}}X_{Z-Z}^{A+1} \quad (2.23)$$

where $T_{1/2}$ - half-life of an unstable isotope $X_{Z'}^{(A+1)}$, Z' - charge carried away by β^- particles; $X_{Z-Z'}^{(A+1)}$- final product of nuclear transformations. An analysis of all possible nuclear reactions makes it possible to predict for each particular material the set of electrically active impurities introduced in this way and their relative activity. i-grade impurity concentration

$$N_{0i}=N_0k_i\sigma_i\varphi t \quad (2.24)$$

where N_0– concentration of the initial mixture of isotopes; k_i and σ_i accordingly, the relative content and activation cross section of the i-th isotope, from which an impurity of the i-th grade is formed; φ – neutron flux density; t– exposure time.

In expression (2.24), the quantities N_0, k_i, and σ_i are constants characterizing the starting material. When irradiated in a nuclear reactor operating at a constant power, the value of φ can also be considered constant. Then the concentration of introduced impurities should depend only on the irradiation time, which can be controlled with a sufficient degree of accuracy; therefore, the set values of the electrophysical properties of the doped materials are quite easily ensured.

However, in order for nuclear doping to prove to be a technically useful way to obtain semiconductor materials with desired properties, the following conditions must be met:

1. the unstable isotopes formed must be short-lived enough for the atoms of the target dopants to form in a sufficiently short period of time after irradiation that is acceptable for technical and economic reasons;

2. the diffusion mean free path of neutrons in the irradiated material should be significantly larger than the size of the doped crystal in order to ensure a fairly uniform distribution of dopants throughout the crystal volume;

3. simultaneously with the target impurities, no other impurity atoms should be formed that worsen the properties of the doped material, or their influence should be small compared with the effect of the target impurities;

4. the influence of impurities present in the source material before irradiation should be either small compared with the target doping effect, or succumb to a fairly strict accounting

5. according to sanitary conditions, the induced radioactivity should be either small or decrease quickly enough to an acceptable level.

Consider from these positions the data on (n,γ) - nuclear reactions for a number of semiconductor materials (table 2.6), compiled using reference data [5] on the absorption and activation cross sections for neutrons with a velocity v=2200 m/s, as well as on half-lives of radioactive isotopes. It follows from the table that reactions 2, 5, 9, and 12 change only the concentration of stable isotopes and do not affect the electrical properties of semiconductors. The remaining reactions lead to the formation of electrically active impurities. In cases where the formation of both donor and acceptor impurities is possible, the final doping result should depend on the relative efficiency of the corresponding reactions, the measure of which is the activation cross section calculated per atom of the initial mixture of isotopes (column 5, table 2.6).

The possibility of changing the electrophysical properties using impurities introduced by the nuclear doping method during neutron irradiation has been proven for all semiconductors listed in Table 2.6. In this case, it was possible to associate certain properties of irradiated semiconductors after appropriate annealing with the nature and concentration of the predominant impurities, in particular, with Ga and As in Ge (annealing for 24 hours at 450°C) [7, 14], with Sn and Te in InSb (annealing for 25–30 h at 620–675°C) [15-17], with Ge and Se in GaAs (annealing at 650–700°C) [18], In in CdS (annealing at 200°C) [19] and with phosphorus in Si [14, 20-24, etc.].

Such a relationship is confirmed, for example, by the coincidence of the carrier concentration measured by the Hall effect with the concentration

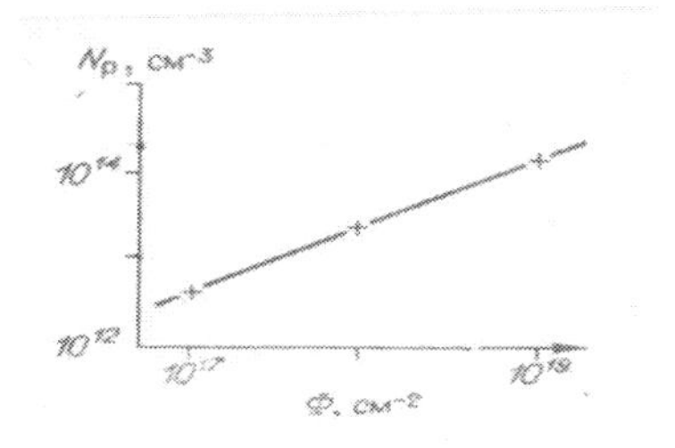

Figure. 2.5. Dependence of the concentration of phosphorus atoms measured by the Hall effect after irradiation in the reactor with various integral neutron fluxes and subsequent annealing at 800 ° for 1 h for p-type silicon with ρ=200 Ohm.cm obtained by by the crucible-free zone melting [23]. The solid line is the phosphorus concentration calculated by the formula (2.24)

of impurities calculated using expression (2.24) for Ge [7], InSb [15], and GaAs [18]. This coincidence is especially characteristic of silicon, in which impurities of only one type (phosphorus) are formed under the influence of slow neutrons. It is shown in fig. 2.5 according to [23, 24], and in fact is the basis of a controlled process of nuclear doping, as a result of which it is possible to introduce a given amount of donor impurities into the *n*- and *p*-type starting material, and in the latter case, doping can be reduced to a change in the type of conductivity. This process, which makes it possible to obtain silicon with a high uniformity of the distribution of dopants by the volume of crystals, has already been carried out commercially in several countries (see, for example, [25–27]).

To assess the nature of the distribution of impurities over the thickness of the doped materials, we will proceed from the fact that the attenuation of the intensity of a collimated neutron beam by a layer of material of thickness x obeys the well-known law

$$I = I_0 \exp(-N\sigma x) = I_0 \exp(-\mu x) = I_0 \exp(-x / l_n) \qquad (2.25)$$

where N - the number of atoms of the irradiated substance in 1 cm^3; l_n- average absorption length [11] associated with the macroscopic absorption coefficient of the material $\mu = N\sigma$ correlation

Table 2.6. Slow neutron nuclear reactions for some semiconductor materials

Material	Isotope	Isotope concentration, %	Absorption or activation cross section, barn		Reactions and end products of nuclear transformations	Impurity effect
			per atom of an isotope	per atom of isotope mixtures		
1	2	3	4	5	6	7
Germanium	Ge	natural		2,45±0,20	1) $Ge_{32}^{70}(n,\gamma)Ge_{32}^{71} \xrightarrow[11{,}4\text{ days}]{K\text{-capture}} Ge_{31}^{71}$	a
	Ge^{70}	20,55	3,42±0,35	0,7	2) $\{Ge^{72}, Ge^{73}\}(n,\gamma) \rightarrow \{Ge^{73}, Ge^{74}\}$	n
	Ge^{72}	27,37	0,98±0,09	0,26	3) $Ge_{32}^{74}(n,\gamma)Ge_{32}^{75} \xrightarrow[82\text{min}]{\beta^-} As_{33}^{75}$	d
	Ge^{73}	7,67	14±1	1,1	4) $Ge_{32}^{76}(n,\gamma)Ge_{32}^{77} \xrightarrow[12{,}1\text{hours}]{\beta^-} As_{33}^{77} \xrightarrow[38{,}7\text{hours}]{\beta^-} Se_{34}^{77}$	d
	Ge^{74}	36,74	0,21±0,08	0,07		
	Ge^{76}	7,67	50±10	3,8±0,8		
Silicon	Si	natural		0,16±0,02	5) $\{Si^{28}, Si^{29}\}(n,\gamma) \rightarrow \{Si^{29}, Si^{30}\}$	n
	Si^{28}	92,18	0,08±0,03	0,07	6) $Si_{14}^{30}(n,\gamma)Si_{14}^{31} \xrightarrow[2.62\text{hours}]{\beta^-} P_{15}^{31}$	d
	Si^{29}	4,71	0,28±0,09	0,01		
	Si^{30}	3,12	0,11±0,01	0,01		
Cadmium sulfide	Cd	natural		2537±9	7) $Cd_{48}^{106}(n,\gamma)Cd_{48}^{107} \xrightarrow[6{,}74\text{ hours}]{K\text{-capture}} Ag_{47}^{107}$	a
	Cd^{106}	1,22	1,0±0,5	0,01	8) $Cd_{48}^{108}(n,\gamma)Cd_{48}^{109} \xrightarrow[453\text{ days}]{K\text{-capture}} Ag_{47}^{109}$	a
	Cd^{108}	0,88	-	-	9) $\{Cd^{110}, Cd^{111}, Cd^{112}, Cd^{113}\}(n,\gamma) \rightarrow$	n
	Cd^{110}	12,39	0,2±0,1	0,02	$\rightarrow \{Cd^{111}, Cd^{112}, Cd^{113}, Cd^{114}\}$	
	Cd^{111}	12,75	-	-	10) $Cd_{48}^{114}(n,\gamma)Cd_{48}^{115} \xrightarrow[53.5\text{ hours}]{\beta^-} In_{49}^{115}$	d
	Cd^{112}	24,07	0,03±0,0015	0,01	11) $Cd_{48}^{116}(n,\gamma)Cd_{48}^{117} \xrightarrow[2.9\text{ hours}]{\beta^-} In_{49}^{117} \xrightarrow[1.93\text{ hours}]{\beta^-} Sn_{50}^{117}$	d

	Cd^{113}	12,26	20 000	12 000		
	Cd^{114}	28,86	1,1±0,3	0,3		
	Cd^{116}	7,58	1,5±0,3	0,1		
	S	natural		0,52±0,02	12) $\{S^{32}, S^{33}\}(n,\gamma) \rightarrow \{S^{33}, S^{34}\}$	n
	S^{32}	95,0	-	-	13) $S_{16}^{34}(n,\gamma)S_{16}^{35} \xrightarrow[87.9\ days]{\beta^-} Cl_{17}^{35}$	d
	S^{33}	0,760	-	-	14) $S_{16}^{36}(n,\gamma)S_{16}^{37} \xrightarrow[5.07\ days]{\beta^-} Cl_{17}^{37}$	d
	S^{34}	4,22	0,26±0,05	0,01		
	S^{36}	0,014	0,14±0,04	0,01		
Antimony indium	In	natural		194±2	15) $In_{49}^{113} \xrightarrow[49\ days]{\beta^-} Sn^{114}$	d
	In^{113}	4,23	56±12	2,5	16) $In_{49}^{115}(n,\gamma)In_{49}^{116} \xrightarrow[54.0\ min]{\beta^-} Sn_{50}^{116}$	d
	In^{115}	95,77	155±10	148		
	Sb	natural		5,7±1,0	17) $Sb_{51}^{121}(n,\gamma)Sb_{51}^{122} \xrightarrow[2.8\ days]{\beta^-} Te_{52}^{122}$	d
	Sb^{121}	57,25	6,8±1,5	3,9	18) $Sb_{51}^{123}(n,\gamma)Sb_{51}^{124} \xrightarrow[60.4 days]{\beta^-} Te_{52}^{124}$	d
	Sb^{123}	42,75	2,5±1,5	1,1		
Gallium arsenide	Ga	natural		2,80±0,13	19) $Ga_{31}^{69}(n,\gamma)Ga_{31}^{70} \xrightarrow[21.1 min]{\beta^-} Ge_{32}^{70}$	d
	Ga^{69}	60,2	1,4±0,31	0,84	20) $Ga_{31}^{71}(n,\gamma)Ga_{31}^{72} \xrightarrow[14.12 hours]{\beta^-} Ge_{32}^{72}$	d
	Ga^{71}	39,8	5,0±0,5	2,0		
	As^{75}	100	5,4±1,0	5,4	21) $As_{33}^{75}(n,\gamma)As_{33}^{76} \xrightarrow[26.4 hours]{\beta^-} Se_{34}^{76}$	d

$$l_n = 1/N\sigma = 1/\mu \tag{2.26}$$

and characterizing the thickness of the layer of material on which the neutron flux density and, accordingly, the concentration of impurities are reduced by $e = 2.72$ times.

The above expressions are valid if we assume that the neutron scattering cross section is small compared with the absorption cross section. In the general case, when there is both absorption and scattering of neutrons, it is necessary to use the conclusions of the general theory of neutron diffusion, according to which (see [11])

$$I = I_0 \exp(-x/L) \tag{2.27}$$

where L - diffusion length, which characterizes, on the one hand, the distance at which the neutron density and concentration of impurities decrease by a factor of e, and on the other, the average distance in a straight line from the beginning to the end of the zigzag neutron trajectory ending at the absorption site.

For substances having atoms of several types, attenuation of the radiation flux can be considered as an additive property of the medium. To account for this fact, we represent expression (2.25) in the form

$$I = I_0 \exp\left(\frac{-\mu}{p} m\right) \tag{2.28}$$

where ρ - substance density; μ/ρ - mass attenuation coefficient; m - mass of a column of substance with a cross section of 1 cm^2 and thickness x. Then for a complex substance

$$\mu/\rho = \sum_i C_i \left(\frac{\mu}{\rho}\right)_i \tag{2.29}$$

where C_i - weight concentration of the i-th element in the mixture.

Expressions (2.25) - (2.29) are also applicable to the case of attenuation of a narrow beam of γ quantum. In this case, the linear attenuation coefficient is determined (see [3]) by the sum of the contributions from the photo and Compton effect, as well as from the process of formation of electron-positron pairs, that is

$$1/\mu = N(\sigma_f + \sigma_k + \sigma pa) \tag{2.30}$$

where σ_f, σ_k, σ_{pa} - cross-sections of these processes of interaction of γ-quantum with matter per atom.

The efficiency of absorption of neutrons and gamma rays by various semiconductor materials under unilateral irradiation is shown in Table 2.7. The μ/ρ values and the corresponding l_n and l_γ values were calculated using expression (2.29), information on the neutron absorption cross sections [5], and extrapolation of the data [3] on the mass absorption coefficients of γ radiation for some elements. The l_n values refer to the energy of γ-quantum of 20-30 MeV, chosen on the basis that it exceeds the threshold energy of the photonuclear (γ,n) reactions for all nuclei shown in tables 2.4 and 2.7.

Table 2.7 also lists the known data on the values of L [5] obtained for Ge and Si taking into account the absorption cross sections σ_n and scattering σ_p, which have values σ_n=2.45±0.2 bar, σ_p=3±1 bar for Ge and σ_n=0.16±0.02 bar; σ_p=1.7+0.3 bar for Si. Unfortunately, there is no data on the value of L for other semiconductor compounds; therefore, a direct comparison of l_γ and L is possible only for Ge and Si, but the general trend can also be judged by the value of l_p.

Table 2.7. The efficiency of attenuation of neutrons and γ-radiation by various semiconductor materials

Alloy	ρ, g/cm	For slow neutrons			For γ rays with E=20-30 MeV		
		$N\sigma$, cm^{-1}	l_n, cm	L, cm	μ/ρ, cm^2/g	μ, cm^{-1}	l_γ, cm
Si	2,42	0,008	125,0	22,2	0,024	0,058	17,2
Ge	5,46	0,25	4,0	4,7	0,039	0,213	4,7
GaAs	5,4	0,36	2,8	-	0,039	0,210	4,8
InSb	5,78	7,0	0,14	-	0,052	0,391	3,3
CdS	4,82	115,0	0,01	-	0,045	0,217	4,6

A comparison of the data given in Table 2.7 shows that for neutrons, as the absorption capacity increases from Si to CdS, the layer thickness of semiconductor materials sharply decreases, within which the distribution of impurities can be considered uniform. Moreover, in Si, the natural inhomogeneity of doping due to absorption and scattering of neutrons is small and does not exceed 1% for real ingots with a diameter of 50 mm [25], and for a crystal with a diameter of 80 mm the ratio of impurity concentrations at the center (C_{min}) and at the periphery (C_{max}) the ingot is Cmin / Cmax ≈ 0.956 [26] (see also chapter 4).

It is of interest to evaluate the uniformity of doping depending on the heterogeneity of the distribution of impurities in the starting material and on the uniformity of the introduction of impurities in the process of doping. If we denote the coefficients of homogeneity of the distribution of impurities in the original (C_{or}) and doped (C_{al}) material from g_{or}= $C_{or}^{min}/C_{or}^{max}$ and g_{al}= $C_{al}^{min}/C_{al}^{max}$, then the degree of doping is characterized by the ratio G_{al}= $C_{al}^{max}/C_{al}^{min}$. Table 2.8 shows the values of g_{al}, characterizing the uniformity of the distribution of impurities (P^{31}) in silicon, depending on g_{or} and g_{al}. These data do not take into account the additional inhomogeneity arising due to the inhomogeneity of the neutron beam in which the doped crystal is located. Therefore, the creation of conditions for uniform irradiation of an doped crystal volume is the main task on the way to obtain silicon with a uniform distribution of phosphorus (see below). It is practically impossible to obtain uniformly doped crystals with a diameter of several centimeters for other semiconductors, but we can talk about doping of wafers cut from the corresponding crystals.

Table 2.8. Uniformity distribution coef. of impurities in doped silicon (g_{al}) depending on the homogeneity of the starting material (g_{or}) and the degree of doping (G_{al})

G_{al}	Values								
	0,1	0,2	0,3	0,4	0,5	0,6	0,7	0,8	0,9
1	0,1	0,2	0,3	0,4	0,5	0,6	0,	0,8	0,9
2	0,55	0,6	0,65	0,7	0,75	0,8	0,85	0,9	0,95
5	0,82	0,84	0,86	0,88	0,9	0,92	0,94	0,96	0,98
7	0,87	0,89	0,90	0,91	0,93	0,94	0,96	0,97	0,99
10	0,91	0,92	0,93	0,94	0,95	0,96	0,97	0,98	0,99
20	0,955	0,96	0,965	0,97	0,975	0,98	0,985	0,99	0,995
50	0,98	0,98	0,99	0,99	0,99	0,99	0,99	0,996	0,998
100	0,991	0,992	0,993	0,994	0,995	0,996	0,997	0,998	0,999

The use of photonuclear reactions opens up additional possibilities in terms of expanding the range of introduced impurities and a higher uniformity of irradiation of materials that are poorly transparent to neutrons. As can be seen from Table 2.7, when compared with neutrons (L), the natural heterogeneity of γ-ray irradiation (l_γ) is approximately the same for Si and Ge and significantly less for other materials.

It should be noted that the large values of L and l_γ for silicon indicate the preference for applying nuclear doping technology to it when using both slow neutrons and γ- radiation. It was indicated above that this has already ensured large-scale doping of silicon using neutron irradiation, and a similar development of silicon doping technology using photonuclear reactions can be expected.

2.6. The influence of side factors on nuclear doping

Nuclear Transmutation by Products

Already from previous reasoning, we can conclude that the specified result of nuclear doping for a particular material, as a rule, cannot be obtained in its pure form, since for each type of radiation and energy of irradiating particles, various nuclear transmutations can simultaneously occur with the formation of a set of products. Therefore, it is desirable to know exactly both the type and effectiveness of all possible reactions, the nature of the influence of the resulting impurities on the properties of the irradiated material (donor or acceptor). Then it is possible to calculate the change in the concentration of charge carriers as a result of irradiation. Unfortunately, this is far from always possible with the necessary accuracy.

For example, in a nuclear reactor, the energy composition of the neutron flux in each specific irradiation site is somewhat different from irradiation in another place and can partially change when materials not related to the doping process are introduced into the reactor. When doping real extended crystals, the geometric parameters of the beam of irradiating particles cannot be considered homogeneous either. It should also be borne in mind that both biographical impurities and impurities obtained as a result of nuclear transmutations can themselves participate in nuclear reactions and produce secondary products of nuclear transmutations. And finally, in the reactor, along with neutron radiation, there is always γ radiation, the maximum energy of which reaches 7.5 MeV [5] and which can create a certain amount of impurities due to photonuclear reactions.

All this requires the most rigorous consideration of these factors when organizing the doping of specific materials. However, in some cases, the total effect of incidental impurities can be estimated theoretically or experimentally and with the help of certain technical methods to eliminate the effect of irregularities of irradiation on the result of doping. Let us consider from these positions the possible role of side nuclear reactions in silicon, in addition to the main (n,γ) reaction with the production of phosphorus. Such adverse reactions can occur at fast neutrons. In particular, the following reactions are possible:

$$Si^{28}(n,2n)Si^{27} \xrightarrow[14.4\,sec]{\beta^+} Al^{27}; \quad Si^{29}(n,2n)Si^{28}; \quad Si^{30}(n,2n)Si^{29};$$

$$Si^{28}(n,p)Al^{28} \xrightarrow[2{,}31\,min]{\beta^-} Si^{28}; \quad Si^{29}(n,p)Al^{29} \xrightarrow[6{,}6\,min]{\beta^-} Si^{29};$$

$$Si^{30}(n,p)Al^{30} \xrightarrow[3{,}3\,sec]{\beta^-} Si^{30};$$

$$Si^{28}(n,\alpha)Mg^{25}; \quad Si^{29}(n,\alpha)Mg^{26}; \quad Si^{30}(n,\alpha)Mg^{27} \xrightarrow[9{,}46\,min]{\beta^-} Al^{27};$$

To assess the contribution of these reactions, it is necessary to know the threshold energy, the activation cross section, and the relative number of neutrons in the spectrum of a reactor with energy above the threshold for each of them. As a side effect, it should also be noted that atoms P^{31} able to undergo further transmutations according to the scheme

$$P^{31}(n,\gamma)P^{32} \xrightarrow[14.3\,days]{\beta^-} S^{32} \qquad (2.31)$$

Since in this case the concentration of P^{31} atoms changes during irradiation, the concentration of S^{32} should also depend on the integral neutron flux as follows: $N_S^{32}=N_p^{31}\sigma_p^{31}\varphi t$, whence $N_S^{32}/N_p^{31}=\sigma_p^{31}\varphi t$. Usually, integrated fluxes of slow neutrons no higher than $\varphi t \approx 10^{19}$ cm^{-2} are sufficient for doping real Si crystals. Then, for the indicated case, taking into account $\sigma_p^{31} = 0.19$ barn [5] $N_S^{32}/N_p^{31} = 0.19 \cdot 10^{-5}$.

Table 2.9 Comparative efficiency and products of nuclear reactions under the influence of neutrons in silicon

Isotope and its concentration, %	Type of reaction	Threshold energy, MeV	Reaction cross section, barn	Half-life $T_{1/2}$	Reaction products	The relative effectiveness of the reactions
Si^{28}, 92,18	n, γ	-	-	-	Si^{29}	-
	$n, 2n$	17,4	10^{-2}-10^{-3}	4,14 sec	Al^{27}	$<10^{-4}$
	n, α	2,7	$\sim10^{-4}$	-	Mg^{25}	$\sim3*10^{-3}$
	n, p	3,9	0,004	2,31 min	Si^{28}	-
Si^{29}, 4,71	n, γ	-		-	Si^{30}	-
	$n, 2n$	8,7		-	Si^{28}	-
	n, α	0,04	10^{-4}	-	Mg^{26}	$\sim10^{-3}$
	n, p	3,2		6,6 min	Si^{29}	-
Si^{30}, 3,12	n, γ	-	0.11	2,62 hours	P^{31}	1,0
	$n, 2n$	>10,0		-	Si^{29}	-
	n, α	4,2	$0.1*10^{-3}$	9,46	Al^{27}	$\sim3*10^{-5}$
	n, p			3,3 sec	Si^{30}	-
P^{31}, 100	n, γ	-	0,19	14,28 days	S^{32}	$<2*10^{-6}$

It is known that the threshold energies $(n, 2n)$ of reactions are related to the threshold of reactions (γ, n) by the relation (see [11])

$$E_{n(n,2n)} = \frac{A+1}{A} E_n(\gamma, n) \tag{2.32}$$

and the quantities $E_n(\gamma,n)$, in turn, are close to the binding energies of neutrons in the nuclei (see table 2.4). It follows that for Si^{28} $E_{n(n,2n)} > 17$ MeV, and since the number of neutrons with such energy in the spectrum of the reactor is very small [28], and the cross sections $(n,2n)$ of reactions for nuclei adjacent to Si are 10^{-2}-10^{-3}barn [5], the amount of impurities Al^{27} formed should be small. The threshold energies of the (n,α) reactions with the formation of Mg^{25} and Mg^{26} are lower, but the cross section of these reactions is apparently very small, because in the case of Si^{30} for neutrons in the fission spectrum it is only 0.1^{-3} barn [29].

Table 2.9 shows the results of evaluating the relative efficiency of all possible nuclear reactions in silicon, taking into account the real energy distribution of neutrons in the spectrum of the WWR-t reactor (Fig. 2.6), as well as data on threshold energies and cross sections of the corresponding reactions [5, 29, 30]. As you can see, some side nuclear reactions affect only the isotopic composition of silicon, and reactions that give other impurities cannot have a noticeable effect on the efficiency of nuclear doping of silicon with phosphorus.

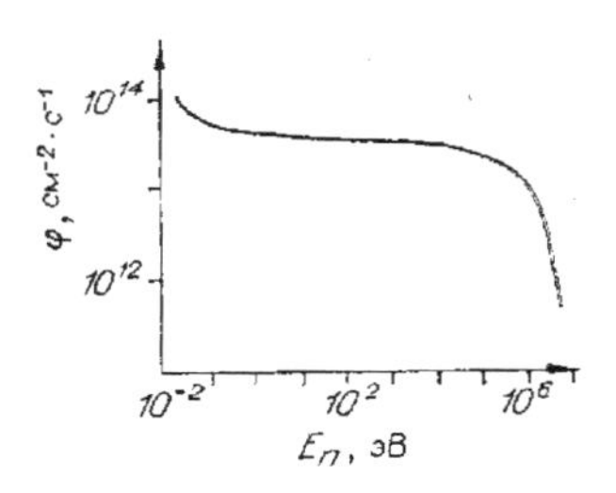

Figure 2.6. The energy spectrum of neutrons in a vertical channel located on the periphery of the active zone of the WWR-t reactor.

Thus, a change in the properties of silicon during nuclear doping can be almost entirely related to phosphorus. The excellent agreement between the measured and calculated carrier concentrations shown in Fig. 2.5, confirms this conclusion on an experimental basis, at least for silicon grown by the crucibleless zone melting method.

A similar analysis can be done for any other specific material subjected to nuclear doping, and we will not dwell on this further.

The radioactivity of doped materials

In many of the nuclear reactions discussed above, first unstable intermediate products are formed that undergo radioactive decay with the emission of one or another type of radiation. In this sense, expression (2.24), strictly speaking, is valid only when, after irradiation, enough time has passed for the complete decay of the intermediate radioactive products. Moreover, during the entire decay time, the doped material will have artificial radioactivity.

When doping silicon, we are dealing with three simultaneously proceeding processes: 1) accumulation of a radioactive isotope Si^{31} in accordance with equation (2.24); 2) decay Si^{31} with education P^{31}, half-life $T_{1/2}$=2,62 h and constant decay λ_1; 3) activation P^{31} and decay P^{32}, half-life $T_{1/2}$=14,3 days and constant decay λ_2. If denoted by N_1 and N_2 respectively, the concentration of atoms Si^{31} and P^{31}, then their change in time due to decay should be described by known differential equations (see [2])

$$\frac{dN_1(t)}{dt} = -\lambda_1 N_1(t); \quad \frac{dN_2(t)}{dt} = \lambda_1 N_1(t) - \lambda_2 N_2(t); \tag{2.33}$$

whose solution has the form

$$N_1(t) = N_{10}\exp(-\lambda_1 t); \tag{2.34}$$

$$N_1(t) = N_{20}\exp(-\lambda_2 t) + \frac{\lambda_1 N_{10}}{\lambda_2 - \lambda_1}\left[\exp(-\lambda_1 t) - \exp(-\lambda_2 t)\right];$$

Similarly to a change in the concentration of radioactive nuclei, the radioactivity due to them should also change, i.e.

$$S_i = S_{i0}\exp(-\lambda_i \cdot t). \tag{2.35}$$

As $\lambda_i N_i$ makes sense of the number of acts of radioactive decay per unit time, it is obvious that $S_{io} = N_{io}\lambda_i$

For simplicity, we first consider the accumulation and decay of any isotope as an independent process. Then if t_{ir} - irradiation time, it follows from (2.24) that $N_{i0} = N_o k_i \sigma_i \varphi t_{ir}$ and therefore:

$$S_i = N_0 k_i \sigma_i \varphi t_{обл} \lambda_i \exp(-\lambda_i t) \tag{2.36}$$

From this expression, it is possible to determine the radioactivity of any product depending on a given exposure time and subsequent decay. In particular, after 3-5 days after irradiation, the specific radioactivity of silicon, due to β-decay in the (n,γ) -reaction of phosphorus production, becomes lower than permissible by sanitary standards [31], which allows working with the material as with stable isotope.

Evaluation of possible sources of radioactivity in silicon associated with the activation of impurities in the starting material shows that almost all possible impurities could create a specific activity equal to the maximum allowable only at concentrations several orders of magnitude higher than their concentrations in real crystals [27].

At sufficiently large irradiation times, a real source of radioactivity appears associated with the reaction (2.31). This activity can also be estimated using (2.36), despite the fact that part of the Si^{31} atoms, the accumulation of which is determined by expression (2.24), decays directly during irradiation. Taking this factor into account, without qualitatively affecting the conclusions, should somewhat reduce the value of activity obtained from (2.36).

Due to the fact that reaction (2.31) is the main source of radioactivity in the production of NDS with specific resistance ρ≤10 Ohm.cm, we use a more rigorous expression for the specific activity S_p of the P^{32} isotope per unit mass of silicon, which depending on the neutron flux density has view [32]

$$S_p = \left(\frac{M_1 k_{Si^{30}} \sigma_{P^{32}} W \varphi^2}{A_1}\right)\left[\frac{\lambda_2}{\lambda_1(\lambda_2-\lambda_1)}\right]\left\{\left[\left(e^{-\lambda_1 N_p/K\varphi}-1\right)+\frac{\lambda_1 N_p}{K\varphi}-\frac{(\lambda_1 N_p/K\varphi)^2}{2}\right]-\frac{\lambda_1}{\lambda_2}\left[\left(e^{-\lambda_2 N_p/K\varphi}-1\right)+\frac{\lambda_2 N_p}{K\varphi}-\frac{(\lambda_2 N_p/K\varphi)^2}{2}\right]\right\} \quad (2.37)$$

where along with the notation used earlier, are introduced: M_1 - mass of irradiated silicon in grams; A_1 - isotope atomic weight Si^{30}; W - Avogadro number; N_p - concentration of phosphorus atoms; K - constant entering into (2.24) in the form

$$t = N_p / K\varphi \quad (2.38)$$

moreover $K = N_0 K_{Si30} \sigma_{si30} = 2{,}06.10^{-4}$.

The calculation results using (2.37) of the specific radio activity of silicon doped with different denominations at different neutron flux densities are shown in Fig. 2.7 [32]. From these data it follows that when obtaining a NDS with ρ>>5 Ohm.cm, the specific activity is lower than permissible regardless of φ. However, even with such specific activity, the total activity of silicon, depending on its total amount, may turn out to be higher than the permissible (10 μCi [31]), which requires the greater exposure after irradiation, the larger the batch of simultaneously processed material (see [25] and Fig. 4.32).

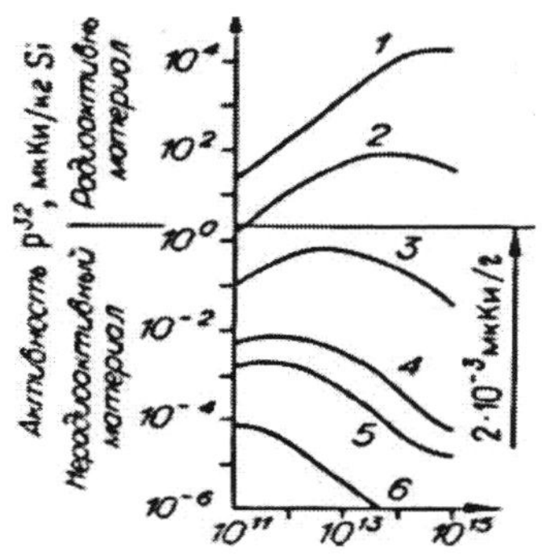

Figure 2.7. The specific activity of the P^{32} isotope, depending on the degree of doping of silicon irradiated with neutrons with different flux densities (specific resistivity, Ohm.cm: 1 - 0.1; 2 - 1; 3 - 10; 4 - 100; 5 - 200; 6 - 1000 [32]).

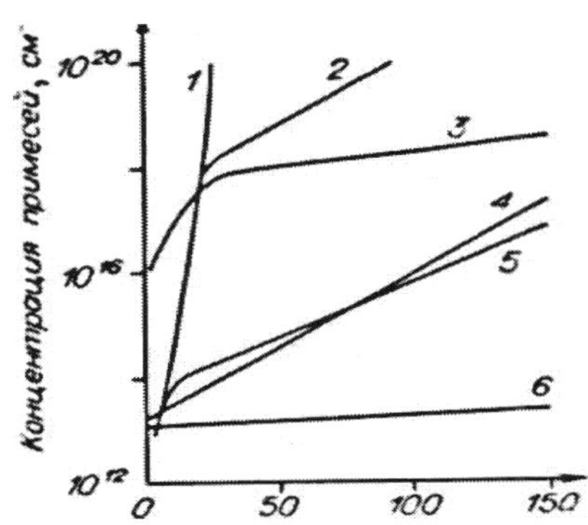

Figure 2.8. The calculated dependences of the decay time to an acceptable level of radioactivity of 1 g of various semiconductor materials on the concentration of impurities introduced by irradiation in the reactor at a thermal neutron flux density of $\varphi=1 \times 10^{14} cm^{-2}s^{-1}$ [32].

1 - GaAs, thermal neutrons; 2 - GaAs, the ratio of thermal to fast neutrons is 10; 3 - Si, 10^{19} Sb/cm^3 atoms; 4 - Ge; 5-GaP; 6-Se.

It should also be noted that when obtaining a NDS with $\rho \sim 10$ Ohm.cm, the specific activity can be reduced by using lower values of φ. This is determined by the fact that the accumulation of P^{32} is proportional to the square φ, however, an increase in the duration of irradiation may prove to be disadvantageous from an economic point of view. The radioactivity of P^{32} also decreases at high values of φ, which is determined by the short irradiation time, insufficient for the complete decomposition of Si^{31} with the formation of P^{31}.

The considered limitations imposed by the radioactivity of neutron-doped materials are the most easily overcome for silicon. For other materials (see table 2.6), the long half-lives of a number of radioactive isotopes make it more difficult to work with irradiated substances under optimal conditions, despite the fact that under the same conditions of irradiation by nuclear doping, one can introduce 100 in germanium and 100 gallium arsenide a concentration of dopants that is 1000 times higher than that of silicon [32]. So, after InSb irradiation with a thermal neutron flux of 2.10^{18} cm^{-2}, the crystal had a large β- and γ-activity, as a result of which it was possible to work with it only after exposure for a year [16, 17].

For a number of other semiconductors in Fig. 2.8 shows the calculated data on the required decay time of radioactivity of 1 g of material to an acceptable level [32]. It can be seen that the situation is relatively more favorable for GaAs, and from the point of view of radioactivity, nuclear doping of this material is better carried out in a reactor with a soft neutron spectrum. At the same time, these data once again emphasize the advantages of nuclear doping primarily for silicon.

Radiation damage

One of the side effects associated with nuclear doping is radiation damage resulting from exposure to various particles or gamma rays. Of greatest practical interest are disturbance formation processes in the volume of irradiated materials, especially when it comes to radiation from a nuclear reactor due to nuclear doping. In this case, the irradiated substance is affected by a neutron flux with a wide energy range, as well as the γ-radiation of the reactor, which, as a result of interaction with the atoms of the irradiated material, can transfer to them enough energy for displacements from the initial positions, and in some cases for secondary displacements under the influence of primary displaced atoms.

If the composition of the radiation is known, then each radiation particle entering the substance forms a unit number of displaced atoms N_{cm} per unit time, moreover,

$$\frac{dN_{cm}}{dt} = N_0 \sigma \Phi \nu_1 \qquad (2.39)$$

where N_0 - concentration of target atoms; Φ - flow of irradiating particles; σ - cross section of their interaction with matter; ν_1 - the number of displaced atoms formed by one particle. In its turn, $\nu_1 = E_t/2E_D$, where E_t - average energy of a moving particle, and E_D - threshold energy of displacement of atoms of the medium.

According to modern concepts (see, for example, 33]), thermal neutrons cannot directly cause atomic displacements due to low energy. However, after such a neutron is captured by a nucleus in (n,γ) - reactions, the intermediate nucleus, as a result of the emission of a γ-quantum according to the law of conservation of momentum, receives a recoil energy equal to:

$$E_{rec} = \frac{E_\gamma}{2Mc^2} = \frac{537}{A} E_\gamma^2 \text{ eV} \qquad (2.40)$$

where: M - weight, A - atomic weight of the recoil nucleus; c - speed of light; $E_\gamma = h\nu$ - energy of the emitted γ-quantum, MeV. Using (2.40) we have

$$\nu_1 = \frac{\bar{E}_{re}}{2E_D} = \frac{1}{4ME_D}\left(\frac{E_\gamma}{c}\right)^2 \qquad (2.41)$$

So, E_{re}= 473 eV, ν_1 = 10 for Si and E_{re} = 186 eV, ν_1= 3 for Ge [33]. According to [34], E_{re} for silicon has an even higher value equal to 780 eV.

β-Decay of unstable products of (*n*, γ) - reactions is also a source of recoil nuclei. In this case, the electron energy E_β according to the laws of relativistic mechanics is connected with its momentum P_β by the relation (see [1, 2])$\sqrt{E_\beta} = \pm\sqrt{c^2P^2_\beta + (m_oc^2)^2}$, where from

$$\bar{E}_{re} = \frac{P_\beta^2}{2M} = \frac{1}{2Mc^2}\left[E_\beta^2 - \left(m_0c^2\right)^2\right] \qquad (2.42)$$

Gamma radiation from radiation capture and β radiation from radioactive decay can be considered as sources of the corresponding “internal” irradiation of the material, which can create additional radiation disturbances. In this case, the electrons create displacements as a result of direct Coulomb interaction with the nuclei, and the γ radiation as a result of the Compton and photoelectric effects, as well as by the creation of electron-positron pairs, also turns into a certain stream of electrons in the substance. The possibility of formation of disturbances by such electrons follows from the fact that for the transfer of energy above the threshold, which for many atoms in semiconductor materials and compounds is contained in the range of 6–16 eV, the energy of the irradiating electrons of the order of 350–380 keV is sufficient (see [35]) In particular, during the decomposition of Si^{31} with the formation of phosphorus, E_β = 1.48 MeV and calculated from (2.42) E_{re} = 32 eV is approximately 2 times higher than E_D.

When irradiated with fast neutrons, in addition to similar effects, direct energy transfer from the neutron to the nuclei as a result of head-on collisions is possible, and from the point of view of the formation of atomic displacements, this process is much more efficient compared to slow neutrons and electrons. In addition, as a result of (n,α), (n,p) and other fast neutron reactions of the irradiated material, a certain number of high-energy charged particles should appear, which also create disturbances in collisions with atoms. However, due to small cross sections (of the order of millibarns) and large threshold energies (see Table 2.1) of the interaction of charged particles with matter, the relative contribution of this mechanism to defect formation can be neglected [36].

Table 2.10 (according to [36]) shows the results of estimates of the number of displaced silicon atoms for each phosphorus atom formed under the influence of various types of radiation when doping silicon in the "hard" (neutron fission spectrum in the reactor core) and "soft" (neutron spectrum in graphite moderator) conditions. It can be seen that the total number of displacements is very large and is mainly due to fast neutrons. Against the background of the contribution of fast neutrons, other sources of radiation disturbances do not play a significant role, but each of them is more effective than the process of formation of phosphorus atoms. The numbers given in table 2.10 should not be regarded as exact values, since they are true only in order of magnitude. This is due to the fact that various radiation defects (vacancies, interstitial atoms, various associations of point defects) can interact with each other and with impurities. Therefore, determining the exact picture of violations by calculation is a very difficult, almost unrealistic task, although many experts consider the number of atomic displacements to be a measure of the absorbed radiation energy. It is believed that the integral neutron flux can serve as a basis for comparing the results of various measurements.

Table 2.10. The number of displaced silicon atoms per atom of phosphorus formed under the influence of various sources of defect formation during doping of silicon in the core and in the graphite moderator of the reactor [36]

Sources of displaced atoms	doping place	
	in the core	in moderator
Fast neutrons	$4{,}06 \cdot 10^6$	$1{,}38 \cdot 10^4$
Gamma quantum fission	$3{,}64 \cdot 10^3$	36,4
Reaction recoil atoms	$1{,}29 \cdot 10^3$	$1{,}29 \cdot 10^3$
β-decay recoil atoms	2,76	2,76
The total number of displacements per atom of phosphorus	$4{,}06 \cdot 10^6$	$1{,}51 \cdot 10^4$

However, due to the fact that the type and efficiency of the formation of defects strongly depend on the neutron energy, the results of irradiation of two identical materials with the same integral neutron fluxes, but with different energy distributions, should differ significantly. To illustrate, in table 2.11 according to [37], it was shown that the efficiency of defect formation in silicon upon irradiation with the same integral neutron flux, but with different average energies, can vary by a factor of 10.

It should also be borne in mind that although defects induced by fast neutrons are formed several hundred times more efficiently than by γ radiation, in some channels of modern research nuclear reactors the flux density of gamma quantum can exceed 100-1000 times the flux density of fast neutrons. As a result, the contribution of γ radiation to defect formation becomes comparable with fast neutrons; therefore, ignoring the contribution of γ radiation can lead to errors of ~ 100% [38]. Some of the displaced atoms, which received energy only slightly above the threshold, can return to their original position even during irradiation.

Table 2.11. Defect formation efficiency in silicon when irradiated with neutrons of different energies

	The cross section for the formation of defects, relative units	
En, MeV	payment	experiment
0,52	0,26	0,24
0,95	0,59	0,55
1,3	0,91	0,9
1,6	1,0	1,0 (taken as a standard)
14	2,46	2,37

Thus, it is difficult and unreliable to theoretically take into account the role of radiation disturbances in the formation of the properties of nuclear doped materials. Therefore, the harmful effect of radiation defects (when their preparation is not an end in themselves) is usually eliminated by appropriate annealing, the temperature and duration of which are determined by the properties of the doped material, the nature, energy composition and intensity of the flow of irradiating particles. Considering the large contribution of fast neutrons to defect formation, one should expect a significant difference in the optimal conditions for defect annealing upon irradiation with the same integral neutron flux, but with a different ratio of the number of thermal and fast neutrons in the spectrum of the reactor (Cd ratio). In this regard, it was shown in [27] that defects in silicon doped in reactors with a ratio of 1000 and 10 at 700°C are sufficiently annealed, and a temperature of ~ 900°C is already needed at a ratio of about 1.

2.7 Possibility of obtaining p-n junctions by nuclear doping

The possibility of introducing impurities by nuclear doping up to a change in the type of conductivity opens up an attractive prospect for the formation of electron-hole transitions directly during irradiation. Indeed, if during the doping process, a part of the p-type irradiated silicon is protected from the action of neutrons and irradiation is carried out until the type of conductivity changes in the unprotected region of the material, then a *p-n* junction should form at the protection boundary. Similarly, one can obtain more complex structures of the type *p-n-p*, *p-i-n*, and others.

Obtaining transitions by the usual method, i.e., by thermal diffusion of impurities at high temperatures, is accompanied by a decrease in the lifetime of charge carriers and other undesirable effects. Nuclear doping has certain advantages due, firstly, to lower transition formation temperatures even in the case when the transition is formed only after annealing of radiation defects (for example, in Si - after annealing at 700-800°C). Secondly, the nuclear method allows, in principle, to obtain a transition of any shape determined by the protection profile, as well as at a given depth of the semiconductor volume.

Experimentally, *p-n* junctions were also obtained by radiation for Ge [21, 39]; however, as indicated, silicon is most suitable for radiation modification of nuclear physical properties, for which the general principles of radiation production of *p-n* junctions, but also analyzed the possibility of implementing these principles in the manufacturing technology of semiconductor devices [21, 40-42] and microcircuits [39].

The successful realization of the possibility of producing electron-hole transitions in real semiconductor devices depends on how much in practice it is possible to reproduce the irradiation conditions that are theoretically necessary for the formation of the transition. These conditions include the effectiveness of protecting (shielding) a part of the crystal from doping, the nature of the directivity of the neutron beam, its energy composition, etc. Therefore, it is necessary to study the influence of all these factors on the efficiency of nuclear doping with each specific radiation source and available shielding means.

We give a general scheme for analyzing the optimal conditions for obtaining electron – hole transitions by the nuclear doping method as applied to silicon [43]. For a qualitatively similar discussion for flat samples, see

[40].

Let a cylindrical Si sample *2r* in diameter be exposed to an isotropic neutron flux through a *2h* wide annular gap in a cylindrical shield of thickness *H* surrounding the sample (Fig. 2.9). If *x* is the distance from the surface of the sample to the point at which a nuclear reaction occurs with the formation of a dopant, and δ is the angle between the *x* axis and the direction of motion of the neutron entering the annular gap, then

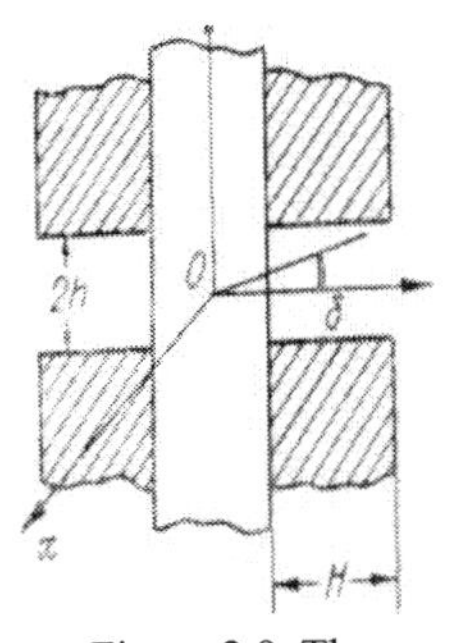

Figure 2.9. The scheme for obtaining the pn junction in a cylindrical sample of a semiconductor placed in an isotropic neutron flux.

$$\delta = \arctan h/(H+x) \qquad (2.43)$$

To simplify the calculations, we find the attenuation of the neutron flux in the middle of the gap on the axis of the sample (point 0 in Fig. 2.9), provided that the protective material is completely opaque and the sample is completely transparent to neutrons. For this case, the possible values of the angle are in the range from 0 to δ_0=arctan *h/(H + r)*.

If in the region where the sample is located, the number of neutrons per unit volume is *n* and there is a Maxwellian distribution of them over the velocities *v*, then in the spherical coordinate system the neutron flux density *dφ* having velocities in the range between v and v + *d*v can be written as (see , for example, [44])

$$d\varphi = n\left(\frac{M_n}{2\pi kT}\right)^{3/2} \exp\left(\frac{M_n \upsilon^2}{2kT}\right)\upsilon^3 \sin\theta d\psi d\upsilon \qquad (2.44)$$

where M_n - neutron mass, T - temperature.

Denoting by φ_g and φ, respectively, the integral density of the neutron flux falling at point 0 of the sample through the protection gap and in the absence of protection, we obtain the attenuation coefficient that determines the collimating effect of the gap:

$$\zeta = \varphi_g / \varphi = \int\limits_{\upsilon=0}^{\infty}\int\limits_{\psi=0}^{2\pi}\int\limits_{\theta=\pi/2-\delta_0}^{\pi/2+\delta_0} d\varphi \Bigg/ \int\limits_0^{\infty}\int\limits_0^{2\pi}\int\limits_0^{\pi} d\varphi = \sin\delta_0 \qquad (2.45)$$

In order to obtain a sufficiently sharp transition, it is necessary to use a narrow gap, which corresponds to small values of δ. In this case

$$\zeta \simeq h(H+r) \simeq h(H+x) \qquad (2.46)$$

Since there are no materials in nature that are completely opaque to neutrons, a certain amount of neutrons will pass through the shield and dope the protected part of the semiconductor. Therefore, if we denote by N_a the concentration of acceptor impurities in the p-type starting material, and by C_{un} and C_{pr}, respectively, the number of dopants formed in a unit volume of the unprotected and protected parts of the sample, then the final values of the carrier concentration in the n- and p- regions of the transition are:

$$N_n = C_{un} - N_a; \quad N_p = N_a - C_{pr} \tag{2.47}$$

It follows that the formation of the p-n junction requires the fulfillment of the condition

$$\xi = C_{pr} / C_{un} < 1 \tag{2.48}$$

To take into account the real energy composition of the neutron flux, we assume that the protective material decreases the total neutron flux due to absorption of neutrons with energy below a certain threshold value in ω pas (in the case of protection from cadmium, ω is the so-called cadmium ratio). Then, for further analysis, it is convenient to represent the value of the C_{un} in

$$C_{un} = C_1 + C_2 + C_3 \tag{2.49}$$

where C_1 and C_2 the number of donor dopants in the unprotected part of the sample formed as a result of nuclear reactions caused by neutrons with energies correspondingly higher below the threshold; C_3 takes into account the additional contribution of neutrons with energies below the threshold, which can enter the unprotected part of the sample through the protective material due to its imperfect absorption capacity. It's obvious that

$$C_1 = \varphi / \omega, \quad C_2 = (\varphi - \varphi / \omega)\zeta \tag{2.50}$$

If σ - neutron absorption cross section, N_0 - the number of nuclei per unit volume of the protection material, and for definiteness, take the path of all neutrons passing through the protection equal to H, then the transmission coefficient of slow neutrons by the protection

$$\gamma = exp(-N_0 \sigma H) \tag{2.51}$$

When

$$C_3 = (\varphi - \varphi / \omega)(1 - \varsigma)\gamma \tag{2.52}$$

As a result

$$C_{нз} = \varphi\left\{(1 - \omega)^{-1}\left[\zeta + (1 - \zeta)\gamma\right] + \omega^{-1}\right\} \tag{2.53}$$

Similarly, for the protected part of the sample, we obtain

$$C_a = \varphi\left[\left(1 - \omega^{-1}\right)\zeta + \omega^{-1}\right] \tag{2.54}$$

and therefore

$$\xi = \frac{\left(1-\omega^{-1}\right)\gamma + \omega^{-1}}{\left(1-\omega^{-1}\right)\left[\zeta + (1-\zeta)\gamma\right] + \omega^{-1}} \tag{2.55}$$

Since it follows from (2.47) and (2.48) that

$$N_P = N_a(1-\xi) - \xi N_n \tag{2.56}$$

then by choosing the appropriate protective material, the geometry of the gap, and the neutron spectrum, one can obtain a predetermined ratio between the carrier concentrations in the *n*- and *p*- regions of the transition for a known concentration of acceptor impurities in the starting material and the required carrier concentration in one of the regions. For specific numerical evaluations, we assume that the protective material has an ideal absorption capacity ($\gamma = 0$), and this cannot significantly affect the results. Then it follows from (2.55) that

$$\omega = 1 + \frac{1-\xi}{\zeta\xi} \tag{2.57}$$

As applied to silicon, we set the initial value N_a=2.10^{15} cm^{-3}, corresponding to the specific resistance ρ=7 Ohm.cm. Further, assuming based on real results, that in the protected part of the sample, under the influence of neutrons with energies above the threshold, the carrier concentration after doping will decrease by no more than 2 times, i.e., N_p=1.10^{15} cm^{-3}, and using (2.56), we find from (2.57) those values of ω that should characterize the neutron spectrum in order to obtain different N_n/N_p ratios in silicon for various given values of the parameter ζ. The calculation results are shown in table 2.12.

As follows from the table, with an increase in the given N_n/N_p ratio, the coefficient ξ decreases, and the required values of the Cd ratio (ω) increase significantly. The obtained values for the cases $N_n/N_p > 5$ at $\zeta=10^{-2}$ significantly exceed the Cd ratio in the channels of WWR type research reactors (ζ~10-20), which are not optimal for obtaining asymmetric electron-hole transitions by the radiation doping method. This qualitatively also explains why, when irradiated in a reactor of a similar type, it is possible to obtain only symmetric pn junctions ($N_n/N_p \sim 1$) [21, 39, 40]. Therefore, a sufficiently effective implementation of the radiation method for producing transitions requires either the creation of specialized reactors with a very soft neutron spectrum (ω>100), or the use of special devices such as a heat column for local "softening" of the spectrum when irradiated in a conventional reactor with a "hard" spectrum.

Table 2.12. The values of the cadmium ratio required to obtain asymmetric p-n junctions in silicon by irradiation in a nuclear reactor under two given conditions for the collimation of a neutron beam ζ

N_n/N_p	ξ	ω	
		$\zeta=10^{-2}$	$\zeta=10^{-1}$
1	$3,0\cdot10^{-1}$	$2,3\cdot10^{2}$	24
2	$2,5\cdot10^{-1}$	$3,0\cdot10^{2}$	30
5	$1,4\cdot10^{-1}$	$6,2\cdot10^{2}$	63
10	$8,3\cdot10^{-2}$	$1,1\cdot10^{3}$	110
100	$1,0\cdot10^{-2}$	$99,9\cdot10^{3}$	990

Improving the effectiveness of protection by reducing its thickness while maintaining good absorption capacity opens up additional opportunities. This, in particular, is achieved by replacing Cd with Gd, using B^{10} and Cd^{113} isotopes instead of a natural mixture of isotopes, and also using combined materials [39, 40].

References:

1. Segre E. Experimental Nuclear Physics. Vol. 1, 2. New York. John Wiley & Sons, Inc. 1955.
2. Mukhin K. N. *Vvedenie v iadernuiu fiziku* [Introduction to nuclear physics]. Moscow. Atomizdat. 1965. 720 p. (in Russian)
3. Gorshkov G. V. *Pronikaiushchie izlucheniia radioaktivnykh izotopov* [Penetrating Radiation of Radioactive Isotopes]. St. Petersburg. Science. 1967. 395 p. (in Russian)
4. Vlasov N. A. *Neitrony* [Neutrons]. Moscow. Science. 1971. 551 p. (in Russian)
5. Kiksin I. K. *Tablitsy fizicheskikh velichin. Spravochnik* [Tables of physical quantities. Directory]. Moscow. Atomizdat. 1976. 1006 p. (in Russian)
6. Price W. J. Nuclear radiation detection. New York-Toronto-London. McGraw-Hill Book company, Inc. 1958.

7. Lark-Gorovits K. *Bombardirovka poluprovodnikov nuklonami* [Semiconductor bombardment by nucleons] in book: Semiconductor materials. 1954, p. 62 - 94 (in Russian)
8. Dolgolenko A. P., Shakhovtsov V. I. *Sozdanie r-n-perekhoda v r-kremnii pod deistviem α-chastits* [Creating a p-n junction in p-silicon under the influence of α particles]. – In book: Radiation Physics of Non-Metallic Crystals. Minsk. Science and Technology. 1970. p. 191-194. (in Russian)
9. Ivanov N. A., Kosmach V. F., Ostroumov V. I. *Imitatsiia radiatsionnogo vozdeistviia neitronov i protonov s pomoshch'iu elektronnykh uskoritelei* [Simulation of the radiation effects of neutrons and protons using electron accelerators]. – In book: Dokl. Second All-Union. conference on the use of charged particle accelerators in the national economy. Vol. 2. St. Petersburg. 1976. p. 198-207. (in Russian)
10. Arifov U. A., Siniukov V. A., Masagutov V. S, Mikhaelian V. M., Korostelev Iu. A., Liutovich A. S. *O vozmozhnosti legirovaniia kremniia s pomoshch'iu fotoiadernykh reaktsii* [On the possibility of doping silicon using photonuclear reactions]. – In book: Crystallization of thin films. Tashkent. Fan. 1970. p. 133 - 135. (in Russian)
11. Curtiss L.F. Introduction to neutron physics. New York-Toronto-London. D. Van Nostrad company, Inc.
12. Cullen D. E., Hlavac P. I. ENDF/B Cross Sections. Brookhaven National Laboratoru, 1972.
13. Brugger R. M., Yelon W. Use of Single Crustal Silicon as a Thermal Neutron Filter.- Proc. Conf. on Neutron Scattering (ed. R. M. Moon), 1976, v. 11, p. 1117-1122.
14. Scheweinler H. C. Some Consequences of Thermal Neutron Capture in Silicon and Germanium.- J. Appl. Phys., 1959, v. 30, № 8, p. 1125-1126.
15. Mirianashvili Sh. M., Nanobashvili D. I., Razmadze 3. G. *O vozmozhnosti transmutatsionnogo legirovaniia antimonida indiia* [On the possibility of transmutation doping of indium antimonide]. Solid state physics. 1965, Vol. 7. № 12. p. 3566-3570. (in Russian)
16. Vodop'ianov L. K., Kurdiani N. I. *Iadernoe legirovanie i*

optichecheskie svoistva sur'mianistogo indiia [Nuclear alloying and optical properties of antimony indium]. Solid state physics. 1966. Vol. 8. № 1, p. 72-76. (in Russian)

17. Vavilov V. S, Vodop'ianov L. K., Kurdiani N. I. *Opticheskie svoistva sur'mianistogo indiia, obluchennogo medlennymi neitronami* [Optical properties of antimony indium irradiated with slow neutrons] – In book: Radiation Physics of Non-Metallic Crystals. Kiev. Naukova dumka. 1976. p. 206-212. (in Russian)
18. Mirianashvili Sh. M., Nanobashvili D. I. *O vozmozhnosti transmutatsionnogo legirovaniia arsenida galliia* [On the possibility of transmutation doping of gallium arsenide]. Semiconductor Physics and Technology. 1970. Vol. 4. № 10. p. 1879-1883. (in Russian)
19. Galushka A. P., Konozenko I. D. *O vliianii reaktornogo izlucheniia na svoistva monokristallov CdS* [On the effect of reactor radiation on the properties of CdS single crystals]. Atomic Energy. 1962. Vol. 13, № 3. p. 277—280. (in Russian)
20. Tanenbaum M., Mills A. D. Preparation of Uniform Resistivity n. Type Silicon by Nuclear Transmutation.— J. Electrochem. Soc, 1961, v. 108, p. 171.
21. Klahr C. N., Cohen M. S. Neutron Transmutation Dope Semiconductors.—Nucleonics, 1964, v. 22, p. 62—65.
22. Arifov U. A., Mikhaelian V. M., Siniukov V. A., Korostelev Iu. A., Liutovich A. S. *Legirovanie kremniia s pomoshch'iu oblucheniia teplovymi neitronami* [Doping of silicon using thermal neutron irradiation] – In book: Crystallization of thin films. Tashkent. Fan. 1970. p. 136-138. (in Russian)
23. Kharchenko V. A., Solov'ev S. P. *Radiatsionnoe legirovanie kremniia* [Radiation Doping of Silicon]. Semiconductor Physics and Technology. 1971. Vol. 5. p. 1641-1642. (in Russian)
24. Kharchenko V. A., Solov'ev S. A. *Radiatsionnoe legirovanie kremniia* [Radiation Doping of Silicon]. Izv. AN SSSR. Neorganich. mater., 1971. Vol. 7. p. 2137-2141. (in Russian)
25. Haas W. E., Schnoller M. S. Silicon Doping by Nuclear Transmutation.- J. of Electronic Materials, v. 5, № 1, p. 57-68.
26. Janus H. M., Malmros 0. Application of Thermal Neutron Irradiation for Large Scale Production of Homogeneous Phosphorus Doping of

Float Zone Silicon.- IEEE. Trans on electrone Devices, 1976, v. ED-23, № 8, p. 797—801.

27. Herzer H. Neutron Transmutation Doping.- In: Semiconductor Silicon, 1977. Eds H. R. Huff, E. Sirtl. Princeton, The Electrochemical Society, p. 106.
28. Neutron Transmutation Doping in Semiconductors. Proc. 2-th Intern. Conf. Columbia, Missouri, April 23-26, 1978. Ed. J. M. Meese. N.Y.- London, Plenum Press, 1979. 371 p.
29. Aliev A. I., Drynkin V. I., Leipunskaia D. I., Kasatkin V. A, *Iaderno-fizicheskie konstanty dlia neitronnogo aktivatsionnogo analiza* [Nuclear Physical Constants for Neutron Activation Analysis]. Moscow. Atomizdat. 1969. (in Russian)
30. Gordeev I. V., Kardashev L. A., Malyshev A. V. *Iaderno-fizicheskie konstanty* [Nuclear Physical Constants]. Moscow. Atomizdat. 1963.
31. *Normy radiatsionnoi bezopasnosti PRB – 76* [radiation safety standards PRB – 76]. Moscow. 1978. (in Russian)
32. Haas E. W., Martin J. A. Nuclear Transmutation Doping from the Viewpoint of Radioactivity.- In: Neutron Transmutation Doping in Semiconductors. Proc. 2-th Intern. Conf. Columbia, Missouri, April 23-26, 1978, Ed. J. M. Meese. N. Y.- London, Plenum Press, 1979, p. 27-36.
33. Kelli B. *Radiatsionnoe povrezhdenie tverdykh tel* [Radiation damage to solids]. Moscow. Atomizdat. 1970. 236 p. (in Russian)
34. Chukichev M. V., Vavilov V. S. *Sredniaia energiia obrazovaniia pary neravnovesnykh nositelei pri obluchenii germaniia γ-luchami So-60* [The average energy of formation of a pair of nonequilibrium carriers upon irradiation of germanium with γ rays of Co-60]. Solid state physics. 1961. Vol. 3. № 3. p. 935 - 942. (in Russian)
35. Vavilov V. S, Ukhin N. A. *Radiatsionnye effekty v poluprovodni¬kakh i poluprovodnikovykh priborakh* [Radiation effects in semiconductors and semiconductor devices]. Moscow. Atomizdat. 1969. 311 p. (in Russian)
36. Meese J. M. The NTD Process - a New Reactor Technology. - In: Neutron Transmutation Doping in Semiconductors. Proc. 2-th Intern. Conf. Columbia, Missouri, April 23-26, 1978, Ed. J. M. Meese. N. Y.- London, Plenum Press, p. 1—10.

37.Serebriakov A. K., Snegirev S. H., Tuturov Iu. F. *Izmerenie potoka neitronov pri radiatsionnykh ispytaniiakh s pomoshch'iu tranzistorov* [Measurement of neutron flux during radiation tests using transistors] – In book: Metrology of neutron radiation at reactors and accelerators. Vol. 2. Moscow. Izd-vo standartov. 1972. p.146-153. (in Russian)

38.Brikman B. A., Kramer-Ageev E. A. *O metodike provedeniia radiatsionnykh ispytanii v reaktorakh* [On the methodology for conducting radiation tests in reactors]. Atomic Energy. 1977. Vol. 42. Iss. 6. p. 461 - 464. (in Russian)

39.Raaen H. P. Fabrication of Semiconductor Devices and Silicon Micro-circuits by Neutron Transmutation Doping. - Isotopes and Radiation Technology, 1970, v. 8, p. 37-60.

40.Ballantine D. Process Radiation Development Isotopes and Radiation Technology. V. 2, 1964-1965.

41.Tanenbaum M. Uniform N-Type Silicon. US patent № 3076732 (1963).

42.German patent № 1214789 (1967).

43.Podsekin A. K., Solov'ev S. P., Kharchenko V. A. *Poluchenie elektronno-dyrochnykh perekhodov metodom radiatsionnogo legirovaniia v iadernom reaktore* [Obtaining electron-hole transitions by radiation doping in a nuclear reactor]. Atomic Energy. 1971. Vol. 31. Iss. 5. p. 521-523. (in Russian)

44.Landau L. D., Lifshits E. M. *Statisticheskaia fizika* [Statistical physics]. Moscow. Science. 1964. 567 p. (in Russian)

Chapter 3 Radiation defects in semiconductors

3.1 Generation of simple radiation defects

This section presents the theoretical calculation of the number of radiation defects that occur in silicon under the influence of electrons, γ-quantum, protons, α-particles, and neutrons.

By radiation defect we mean the simplest defect - the Frenkel pair - vacancy + interstitial atom. The calculations do not take into account that part of the vacancies and interstitial atoms annihilate and form various complexes. Therefore, the calculated concentrations of the simplest defects should be considered as the number of Frenkel pairs introduced by the bombarding particle without taking into account their further "fate".

Electrons. If an electron with energy E elastically scattered by the core of the crystal lattice atom at an angle θ_e from the original direction, then the energy transferred to the atom,

$$\mathcal{E} = \mathcal{E}_{max} sin^2(\theta_e / 2) \tag{3.1}$$

where

$$\mathcal{E}_{max} = (560,8 / A)(E / mc^2)(2 + E / mc^2) \tag{3.2}$$

- maximum energy an atom receives in a head-on collision $\theta_e = \pi$. Since the mass of the core A many times exceeds the mass of the electron, the coordinate system of the center of mass practically coincides with the laboratory. In this case, the angle θ_A at which the recoil atom is scattered is related to the angle θ_e simple ratio

$$\theta_A = (\pi - \theta_e)/2 \tag{3.3}$$

taking into account (3.3), formula (3.1) can be written in the form

$$\mathcal{E} = \mathcal{E}_{max} cos^2\theta_A \tag{3.4}$$

To calculate the number of displaced atoms, it is necessary to know the scattering cross section of incident electrons by nuclei. An approximate solution of the Dirac equation, made by McKinley and Feshbach [1], gives for the differential cross section for scattering of relativistic electrons on light nuclei the expression

$$\begin{aligned} d\sigma = \pi(ze^2 / mc^2)^2 \left[(1-\beta^2)/\beta^4\right]\{1 - \beta^2 \sin^2(\theta_e/2) + \\ + \pi\alpha\beta\left[sin\,(\theta_e/2) - sin^2(\theta_e/2)\right]\}\left[\cos(\theta_e/2)/\sin^3(\theta_e/2)\right] d\theta_e \end{aligned} \tag{3.5}$$

where $\alpha = z/137$; $\beta = v/c$; v - electron speed; c - speed light; z - core charge; m - electron mass; e - electron charge. The obtained expression for elements with a small value of z serves as a good approximation of the Mott formula [21], which describes scattering processes most accurately.

Using formula (3.1), the differential cross section can be expressed as a function of the energy transferred to the atom

$$d\sigma(\mathcal{E}) = \pi\left(ze^2/mc^2\right)^2\left[(1-\beta^2)/\beta^4\right]\left[1-\beta^2\mathcal{E}/\mathcal{E}_{max} + \right.$$
$$\left. +\pi\alpha\beta\left\{(\mathcal{E}/\mathcal{E}_{max})^{1/2} - \mathcal{E}/\mathcal{E}_{max}\right\}\right](\mathcal{E}_{max}/\mathcal{E})^2 d\mathcal{E} \quad (3.6)$$

The complete cross section for the formation of primary displaced atoms is obtained by integrating the last expression in the range from $\mathcal{E}_d$ to $\mathcal{E}_{max}$:

$$\sigma_d = \pi\left(\{ze^2/mc^2\right)^2\left[(1-\beta^2)/\beta^4(\mathcal{E}_{max}/\mathcal{E}_d - 1) - \right.$$
$$\left. -\beta^2 ln(\mathcal{E}_{max}/\mathcal{E}_d) + \pi\alpha\beta\left\{2\left[(\mathcal{E}_{max}/\mathcal{E}_d)^{1/2} - 1\right] - ln(\mathcal{E}_{max}/\mathcal{E}_d)\right\}\right] \quad (3.7)$$

where $\mathcal{E}_d$ - defect threshold energy.

The number of displaced atoms as a result of the bombardment can be relatively easily calculated, assuming that the threshold energy is isotropic:

$$N_D = v\Phi \times n(E_0) \quad (3.8)$$

where Φ - integral electron flux; v - value taking into account the cascade nature of displacements [3]. J. Kinchin and R. Pease [3] value v determined as follows:

$$\begin{aligned} v(\mathcal{E}) &= 1 && at \quad \mathcal{E}_d < \mathcal{E} < 2\mathcal{E}_d \\ v(\mathcal{E}) &= \mathcal{E}/2\mathcal{E}_d && at \quad 2\mathcal{E}_d < \mathcal{E} < \mathcal{E}_1, \\ v(\mathcal{E}) &= \mathcal{E}_1/2\mathcal{E}_d && at \quad \mathcal{E} > \mathcal{E}_1, \end{aligned} \quad (3.9)$$

where $\mathcal{E}_1 = A_1\mathcal{E}_g/8n$, A_1 is the mass of the atom; $\mathcal{E}_g$ is the lowest electron excitation energy, which coincides with the "optical" band gap; $n(E_0) = N\int_0^x d\sigma[E(x)]dx$ - average number of initially displaced atoms created by one electron with initial energy E_0. Here x_0 - effective range of an electron; N - number of atoms in 1 cm^3.

The dependences $n(E_0)$ for germanium and silicon are shown in Fig. 3.1. In the calculation, the value of v was taken into account. A similar dependence for silicon at E_0 up to 100 MeV is shown in Fig. 3.2

γ-quantum. The main role in creating defects is played by the action on the crystal of fast electrons arising as a result of the photoelectric effect and the Compton effect, as well as pairs of electrons and positrons that appear at sufficiently high γ-ray energies. The total absorption cross section for gamma rays with energies up to 10 MeV is determined by the three processes mentioned above. Expressions for the cross section for the formation of a primary displaced atom can be found, for example, in [6].

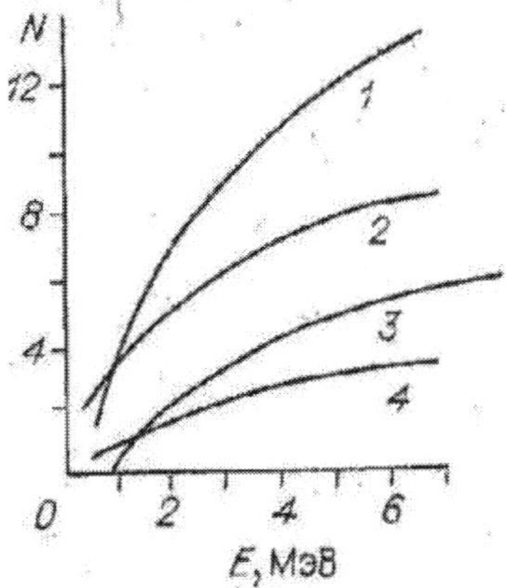

Figure 3.1. The number of displaced atoms per incident one fast electron. E_d = 15 eV: 1 - germanium, 2 - silicon; E_d = 30 eV: 3 - germanium, 4 - silicon.

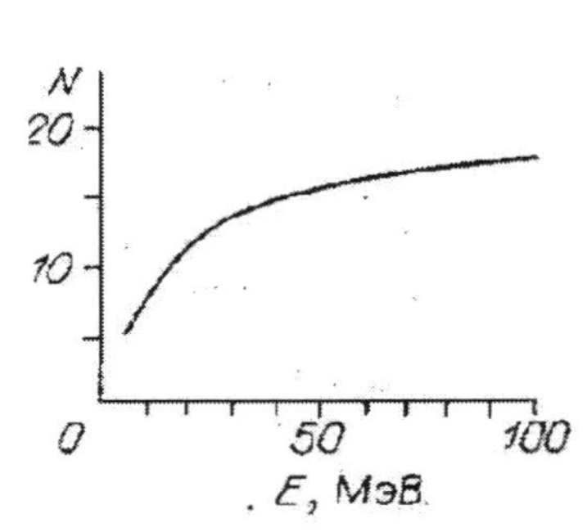

Figure 3.2. The number of simple defects formed by an electron with an energy E in 1 cm^3 of silicon [5].

As an example in fig. Figure 3.3, a, b shows the dependences of the absorption cross section of γ-quantum for germanium and silicon on the photon energy.

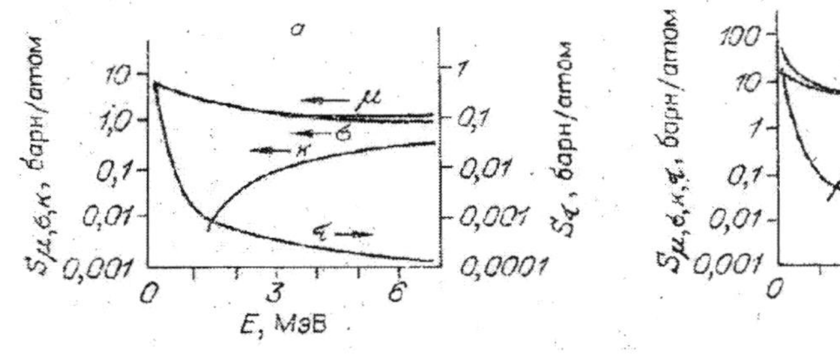

Figure 3.3: Cross section for the absorption of gamma rays for silicon (a) and germanium (b). τ - is the photoelectric effect; σ - is the Compton effect; k - the formation of electron-positron pairs; μ - is the total cross section.

To calculate the concentration of displaced atoms, it is necessary to use the above-described scheme for the appearance of defects under the action of fast electrons. The number of displaced atoms resulting from the action of γ quantum in silicon and germanium was calculated in [4]. In the calculations, the accepted assumption about the existence of a certain threshold energy $\mathcal{E}_d$ was used. The results of calculations [4] are shown in Fig. 3.4.

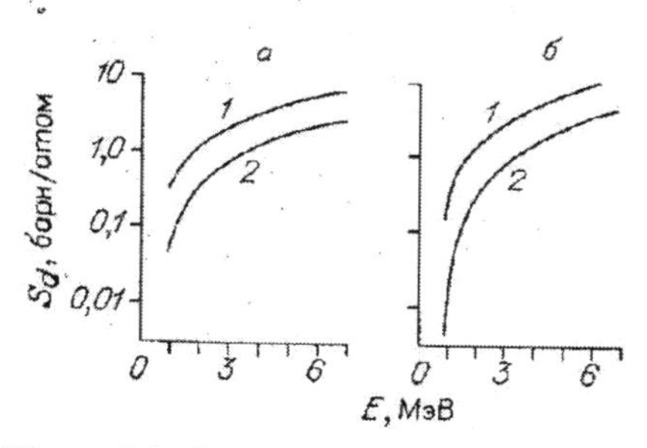

Figure. 3.4. Cross section of the process of displacement of silicon atoms (a) and germanium (b) under the influence of γ-quantum. E_d, eV: 1 - 15; 2 - 30.

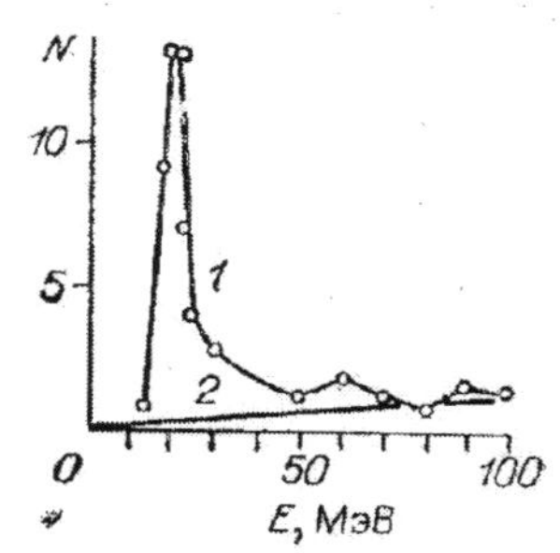

Figure 3.5. The number of simple "nuclear" (1) and "cascade" (2) defects formed by a γ-ray with an energy E in 1 cm^3 of silicon [5].

Thus, the question of the action of γ quantum on semiconductors reduces to calculating the probability of the appearance of electrons and positrons in a crystal. However, such an approach is valid only at γ-ray energies not exceeding 10 MeV. At high energies, the contribution of photonuclear reactions must be taken into account. In [5], the defect concentration was calculated taking into account: ionization losses, bremsstrahlung, electron production, multiple Coulomb scattering of electrons, photoelectric effect, Compton scattering of γ-quantum, formation of electron-positron pairs, annihilation of positrons, elastic scattering of γ-quantum by nuclei, photonuclear reactions. The calculation results for silicon are shown in Fig. 3.5. At γ-ray energies of 20-30 MeV, a giant dipole resonance of the cross section for photonuclear reactions appears, which leads to a sharp increase in the concentration of defects!

Charged heavy particles. At high energies, an atom moving in a solid loses part of its electrons and becomes repeatedly ionized. The degree of ionization or the effective charge of an ion is a complex function of its energy and increases with increasing particle energy. The main reason for energy loss is the excitation of electrons (ionization losses), but sometimes a moving atom can interact directly with the lattice atom. With a decrease in the kinetic energy of a moving atom, the degree of its ionization decreases, and with it the rate of energy loss by excitation of electrons. In the end, a moving atom becomes neutral, and its energy is consumed mainly in collisions with atoms of a solid, which can now be considered solid balls. Thus, irradiation with heavy charged particles leads to a substantially inhomogeneous distribution of displaced atoms.

The results of the calculation in [8] of the total number of defects introduced by protons before they stop completely are presented in Fig. 3.6. The issues related to the effect of protons on germanium and silicon are described in detail in [9]. It also discusses the role of nuclear reactions as a mechanism of radiation damage at high particle energies.

Neutrons. There are two different mechanisms of atomic displacements in solids irradiated with neutrons. In accordance with the first, a neutron collides directly with the nucleus of an atom and transfers to it energy exceeding the threshold. Almost always, the core receives enough energy to form the following secondary displacements. The second mechanism is due to nuclear reactions on neutrons, which are discussed in detail in Sec. 2. The reaction products cause atomic displacements. The results of calculating the average concentration of defects in silicon as a function of the neutron energy carried out by the authors of [10] are presented in Fig. 3.7.

The authors point out the importance of taking into account the contributions of inelastic neutron scattering and the nuclear reactions caused by them to the overall radiation effect.

In conclusion, we emphasize once again that, as a rule, the concentration of displaced atoms or Frenkel defects formed in the “primary” processes is calculated. In fact, one has to deal with a whole spectrum of radiation defects, the type and concentration of which are determined by "secondary" processes: diffusion, annihilation, complexation, coagulation, etc.

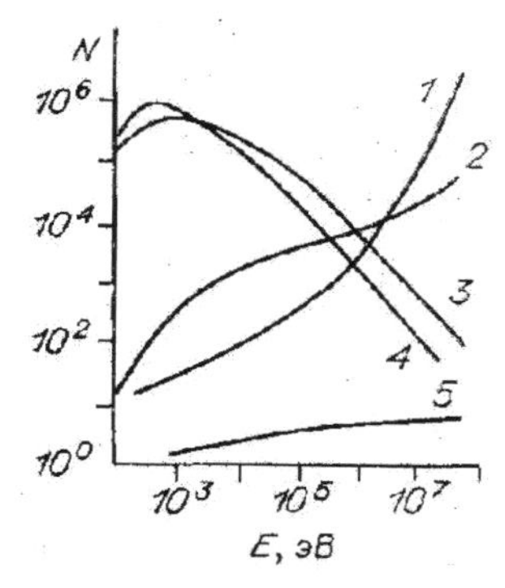

Figure 3.6. Protons in silicon [8]. 1 - path dependence, μg/cm^2, on energy; 2- the total number of defects (multiplied by 100); 3 - defect introduction rate, cm^{-1}; 4 - atomic displacement cross section $\sigma_d \cdot 10^{23}$ cm^2; 5 - defect propagation coefficient.

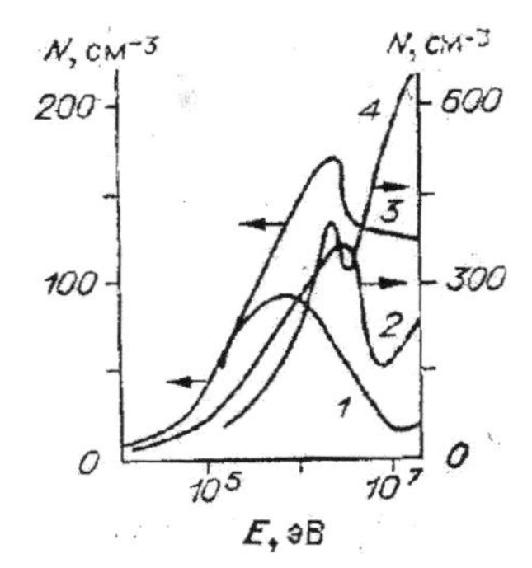

Figure. 3.7. Dependence of the average concentration of defects in a silicon crystal on the neutron energy [10]. 1, 2 - Kramer – Ageev calculations (see [10]) according to the Kinchin - Pease and Lynhard theory; 3 - calculations under the Kinchin – Pease theory; 4 - Lynhard calculations.

3.2 Types of radiation defects in silicon

The largest number of radiation defects identified in silicon. A description of most of them can be found in conference proceedings (for example, [11-15]), as well as in monographs [7, 9, 16-20]. We confine ourselves to listing the main radiation defects and a brief characteristic concerning the charge state of the defect, the temperature region of its existence, and the levels that the associations under consideration introduce into the band gap of silicon.

Vacancy defects. The simplest of these is vacancy. It can be in four charge states: doubly and once negatively charged, neutral and once positively charged, mobile in n-type silicon at $T \geq 60$ K (negatively charged), and in p-type silicon at $T \geq 160$ K (neutral).

In four charge states, divacancy is also observed [21-24]. It can be introduced as a primary defect when irradiated with heavy particles, neutrons or electrons with high energy (> 1 MeV). Divacancies are annealed at temperatures of 520–570 K. At high radiation doses and in dislocation-free materials, the annealing temperature increases. Levels are introduced into the forbidden zone of divacancies E_c - 0,4 eV, E_v + 0,27 eV and level near the middle of the forbidden zone.

A complex of three vacancies arises as a result of annealing of irradiated silicon at T> 420 K and disappears at T> 550 K [15, 25]. Associations of four vacancies come in two configurations: planar and non-planar [22, 26-28].

Defects of this type exist in two charge states: neutral (Si – *P*3 [26]) and singly negatively charged (Si – *S*2 [28]). The EPR signal spectrum of Si-*P*3 disappears at about 450 K. At the same temperatures, the appearance of nonplanar tetravacancy is observed, which is annealed at T ≥ 570 K.

The five-vacancy complex (Si-*P*1) arises upon irradiation of silicon with neutrons and ions [29–31]. It is observed in a once negatively charged state. It arises after annealing at T ≥ 400 K and is stable up to temperatures of ~ 720 K.

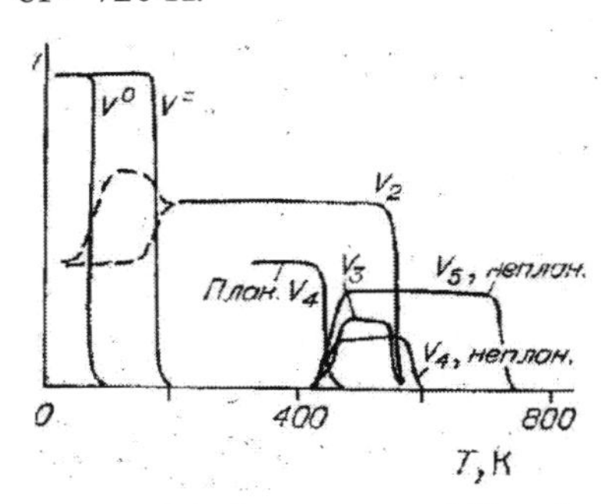

Figure 3.8. Temperature stability of complexes of vacancies in silicon

In [25], based on the analysis of EPR spectra, it is assumed that there are complexes containing a larger number of vacancies. In fig. Figure 3.8 schematically depicts the stability regions of vacancy complexes.

Vacancy complexes with impurity atoms. In irradiated n-type silicon crystals obtained by zone melting in vacuum, EPR signals associated with the vacancy + donor impurity complex (E-center) were detected.

The first model was constructed for the vacancy + phosphorus complex [32]. Then, the E centers were found in silicon doped with arsenic and antimony, and the acceptor level E_c – (0.43 ± 0.03) eV belongs to the E-center. The rate of complex introduction increases with an increase in the covalent radius of the impurity [33]. The annealing temperature of the E-center depends on the type of donor and is 400, 440, and 460 K, respectively, for the donor impurity of phosphorus, arsenic, and antimony. The activation energy of the annealing of the vacancy + phosphorus complex is 0.96 eV. Usually, annealing is interpreted as thermal dissociation of the complex, but the authors of [34] believe that the complex diffuses as a single formation to the drain. Complexes consisting of a vacancy and an impurity of group III were found [35, 36], but these defects are not as well studied as E-centers. The EPR spectra identified vacancy + tin complexes [37], vacancy + germanium [38].

The first identified complex in irradiated silicon is the vacancy + oxygen association (A-center) [39, 40]. This complex introduces an acceptor level of E_c-0.17 eV into the forbidden zone. A defect is observed in silicon of n- and p-types of conductivity at temperatures when vacancies are mobile. The center is stable up to temperatures of 600–650 K. To date, low-temperature modifications of the A-center have been identified, as well as complexes comprising more than one oxygen atom or more than one vacancy [28, 41–43]. As a rule, the more complex the complex, the higher it is annealed at a higher temperature (Fig. 3.9).

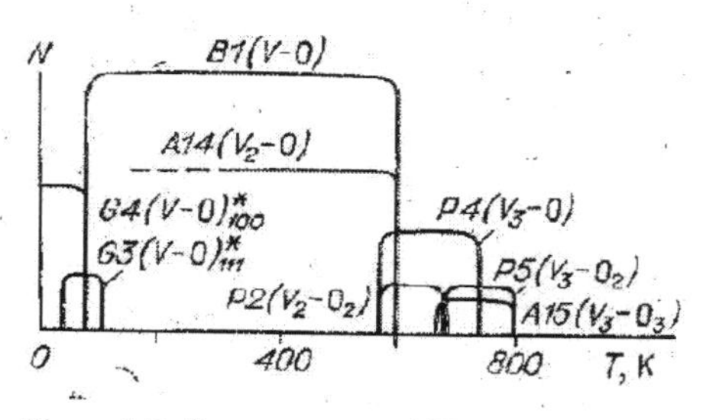

Figure 3.9. Temperature stability of vacancy - oxygen complexes in silicon.

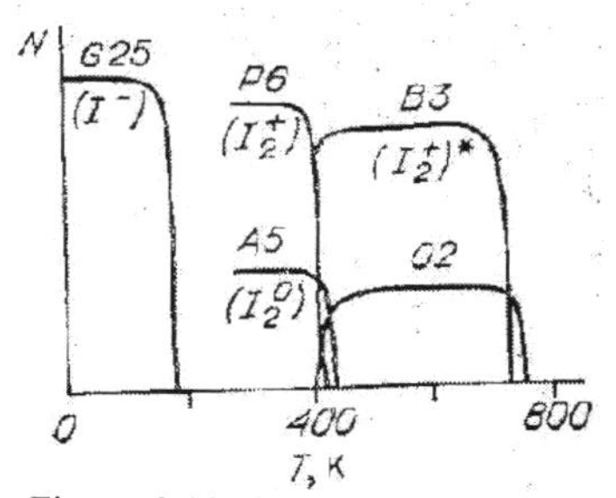

Figure. 3.10. Temperature stability of defects from interstitial silicon atoms.

In p-type silicon, the K-center is the dominant defect. It is known that this defect includes vacancies and oxygen. Only recently has a K-center model been proposed. It consists of carbon + oxygen + divacancies [15].

Thus, if you look at the periodic table, it turns out that vacancy complexes with all the closest "neighbors" of silicon are fixed.

Interstitial defects. In explaining the first experiments [44, 45], it was assumed that if irradiated at a sufficiently low temperature, then the simplest defects would be "frozen". Then raising the temperature, you can fix the beginning of the movement of the defect. These ideas turned out to be true for the vacancy. As for interstitial atoms, all researchers came to a unanimous opinion: they are mobile, at least during irradiation, at very low temperatures up to 0.5 K [46]. All identified complexes belonged to vacancy complexes. The question arose - where are the interstitial atoms?

Apparently, the most definite discussion of the actual interstitial defects in silicon was made after interpreting the results on internal friction in silicon implanted with boron [47–49]. The authors believe that interstitial silicon atoms can be in three charge states: once negatively charged when the Fermi level is above E_c - 0.4 eV, once positively charged when the Fermi level is below E_p + 0.4 eV, and in neutral. The migration energy depends on the charge state and is ~ 0.35; 0.85; 1.5 eV [15], respectively, for negatively, positively charged, and neutral internodes. Around the same time, works appeared that reported the decoding of the spectra belonging to interstitial defects — these are primarily double interstitial atoms (centers *P*6, *B*3, *A*5) [50-54]. The described interstitial defects have different temperature stability (Fig. 3.10).

Of the impurity interstitial defects, the EPR signal spectrum associated with the interstitial aluminum Al++ was the first to decipher [40, 55]. The noted defects are stable up to 200°C. Heating above 200°C leads to a decrease in the concentration of interstitial aluminum and the appearance of defects Al_3 + Al_i - (aluminum in the node + aluminum in the internodes). Qualitatively, the same picture is observed upon irradiation of silicon doped with boron and gallium. The position of the energy levels of the interstitial impurity is unknown.

In conclusion, we consider another impurity — carbon, which is almost always contained in silicon and is easily transferred to the interstitial position by irradiation. In the interstitial position, carbon atoms are already moving near room temperature (~ 40°C) [56, 57]. In crystals with carbon, the rate of introduction of divacancies sharply increases [56]. In the presence of oxygen, this effect is less pronounced. The authors of [56] observed a series of bands that were associated with various complexes, including oxygen and carbon. They have different temperature stability: occur at 160-180°C, anneal at 220-340°C. In silicon with carbon and oxygen, more complex centers are formed. Thus, the authors of [58] interpret the EPR spectrum of Si - *G*15 as coming from the association of [CO + V_2] - the "molecule" of CO and divacancies and ascribe the level of E_v + 0.32 to it. In the same work, it is assumed that the EPR spectrum of G16 is associated with a negatively charged state of the same defect (CO + V_2), which is associated with an E_c-0.43 eV level.

Disordered areas. Irradiation of silicon by electrons with an energy of about 10 MeV, neutrons, and ions gives rise to disordered regions. In the first model [59], it was assumed that the core of the disordered region is a material obtained after irradiation of the semiconductor with extremely large doses and is separated from the undamaged crystal by the space charge region. For a long time, this model was used to explain the experimental data. However, not all results were explained with its help. In 1972, N. A. Ukhin [60] proposed a model for the formation and annealing of disordered regions, taking into account new experimental facts. The essence of the model is as follows.

As a result of the initial act of transferring a sufficiently high energy material to the atom (of the order of several tens of kilo electron-volts), a bunch of monovacancies and interstitial atoms forms. Interstitial atoms, migrating after their appearance, can annihilate with mono-vacancies or leave the damaged area, diffusing into the matrix. Mono-vacancies in addition to these processes can form immobile and stable complexes at a given temperature with impurities and divacancies.

At room temperature, part of the vacancy – interstitial pairs is annealed already for times of the order of 10^{-6} s, when interstitial atoms migrate intensively against the background of practically motionless vacancies. At the same time, due to the large gradient in the concentration of interstitial atoms at the interface between the damaged regions and the main material, the matrix, most of them leave these regions in the indicated time, as a result of which a bunch of monovacancies remain in the damaged region.

In later times (of the order of a few milliseconds), only mono-vacancies migrate within the damaged areas and their immediate surroundings. In this case, divacancies are formed inside the regions, and vacancy complexes with impurities present in high concentrations, for example, with oxygen, form on their periphery. The penetration depth of monovacancies into the matrix and, consequently, the final size of the disordered regions will be determined by the concentration of impurities in the matrix with which monovacancies can form stable complexes.

Thus, the disordered region is a complex formation consisting of a core and a shell saturated with divacancies, including monovacancy complexes with impurities. The formation of quasimolecules $Si0_4$ or SiC is most likely.

High-temperature annealing of such damages should proceed in several stages and be completed at sufficiently high temperatures of 700-800°C. If the assumption of quasimolecules is true, then at the early stage of annealing (200–300°C), divacancies in the nucleus should dissociate with the formation of either additional quasimolecules or vacancy complexes with impurities stable at the indicated temperature. After this annealing stage is completed, only one shell remains in the material, the thickness of which can increase. With further annealing, changes should be observed, associated mainly with the restructuring of these shells up to their complete disintegration.

The N. A. Ukhin model, which is a big step forward, considered disordered regions as systems with constant properties after stabilization. Further development of the model went along the path of concretization of the primary processes of formation and stabilization of disordering regions. Two mechanisms were discussed in the literature, the separation of vacancies and interstitial atoms — the so-called spontaneous inside-cascade annealing of closely spaced pairs [61, 62] and diffusion, which takes into account the large difference in the diffusion coefficients of vacancies and interstitial atoms (see, for example, [63]). In both cases, as shown in the cited works, a very small part (~ 10%) of the defects that initially formed in the cascade "survive".

In [64], not excluding the possible contribution of the above mechanisms, it was suggested that the main mechanism leading to the escape of interstitial atoms from the cascade region is the process of point (or linear) "microexplosion": instantaneous values of the energy released inside the cascade that is effective inside the cascade temperature and effective inside the cascade pressure, they completely provide a quick ejection of a significant part of the displaced atoms outside the cascade (without significant heating and structural damage to the adjacent region of the crystal). The disordered regions finally formed, as a rule, consist of a nucleus, the basis of which are vacancy clusters of various complexity and defective impurity shells arising from the interaction of disordered regions with the environment [65].

The authors of [64, 65] present the formation of a defective impurity shell as follows: interstitial atoms of the main substance, emanating from a disordered region, as a result of interaction with impurities "push" them into interstitial positions. Interstitial impurity atoms, having large diffusion coefficients, diffuse along the crystal mainly in the direction of the disordered region. During diffusion, interstitial impurity atoms, when meeting with a vacancy, form a substitutional impurity, and when meeting with a nodal impurity, they form complexes of the type of interstitial impurity + nodal impurity. Thus, the disordered region is surrounded by a defective impurity shell

The authors of [65] believe that the formation of disordered regions does not end there, especially if irradiation with particles continues and vacancies and interstitial atoms are generated in the crystal. The heterogeneities in the structure of disordered regions, local elastic stresses, and the electric field stimulate their "accommodation" processes to the environment, decoration by point defects and the formation of large complexes, as a result of which, the authors believe, the charge of the disordered region would be close to zero if the properties of the medium remained unchanged for quite a long time. However, under conditions of continued irradiation, the previously disordered regions formed are in a medium whose properties are continuously changing, thereby violating the neutrality of the disordered regions. The processes of "accommodation" tend to restore equilibrium, but since the matrix in this case plays an active role, and disordered regions due to internal stability have some "inertness", changes in their properties will always lag behind changes in the matrix and disordered regions will be charged. From these general considerations, the following conclusions are made regarding the charge of disordered regions:

1. The magnitude of the charge of disordered regions depends on the experimental conditions and the starting material (on how easily and completely the processes of interaction of disordered regions with the environment go).
2. The sign of the charge of disordered regions is determined by the direction of motion of the Fermi level in the matrix during irradiation and the initial state of the material. If the Fermi level decreases, then the charge will be positive, if it increases, then negative.

So, disordered areas will be: a) positive if the n-type starting material and electrical conductivity tend to their own value when the dose is set; b) negative if the p-type starting material and conductivity also tend to an eigenvalue upon irradiation; c) neutral if the conductivity of the starting material is close to its own value and remains intrinsic to radiation.

In [65], it was emphasized that in most of the cases considered, the signs of the charge of the disordered regions and the main charge carriers are opposite, that is, the interaction should be described by the attractive potential. Therefore, the scattering of charge carriers on disordered regions can in some cases have a resonant character, and the influence of disordered regions on the electrophysical properties of crystals can be significant.

3.3 Accumulation of radiation defects

The irradiation of crystals of a semiconductor material leads to a change in its characteristics. The degree of deviation of the properties of an irradiated crystal from an unirradiated one depends on many factors. We will name only the most important: 1- type and energy of the bombarding particles; 2- impurity and defective composition of the irradiated crystal; 3- irradiation conditions (sample temperature, flow rate); 4- interaction of the introduced defects with the initial violations of the crystal lattice; 5- competition for the capture of generated simple defects (vacancies and interstitial atoms) by various violations of the crystal lattice. Under real conditions, changes in the characteristics of a material are determined by a set of secondary radiation defects (see, for example, [19]). In oxygen-containing silicon, the dominant secondary defects are A-centers — the vacancy + oxygen complex. In the dislocation-free material at the initial stages of irradiation, E centers dominate (vacancy + donor impurity atom). The concentration of complexes including impurity tends to saturation. There are two reasons for this saturation. The first, trivial is the depletion of impurities. The second is as follows: as the complexes accumulate, the probability of interaction between the vacancy (or interstitial atom) and these complexes increases with the formation of a more complex association if the same component of the Frenkel pair approaches, or with the release of the impurity if the opposite component is suitable. In the latter case, the Frenkel pair is annihilated and the impurity acts as the center of annihilation.

It is clear that the interactions of primary defects with different impurities will be different. In this case, a certain role is played by the irradiation temperature. The influence of the irradiation temperature on complexation was considered in [19], where there is a detailed list of original works. The main result of these studies is as follows: irradiation at elevated temperatures leads to a change in the spectrum of defects, to the disappearance of those defects whose annealing temperature does not exceed the irradiation temperature. We present the results of experiments indicating that the indicated approach to the role of the irradiation temperature is simplified and requires further research.

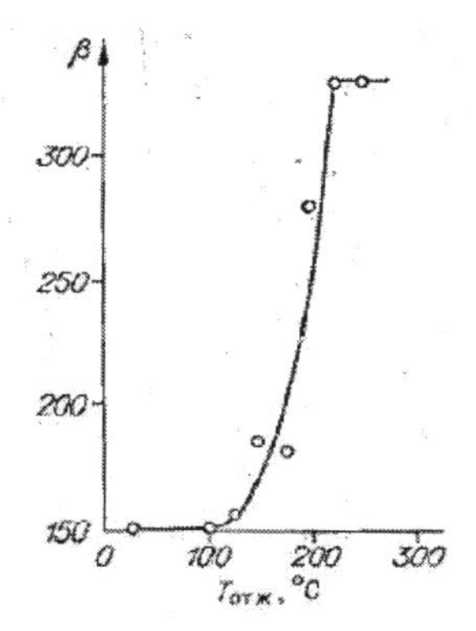

Figure 3.11. Isochronous annealing of silicon transistors irradiated with electrons. T_{area}=150°C, Φ = $8\cdot10^{15}$ el/cm^2; annealing for 15 min; electron energy 3.5 MeV.

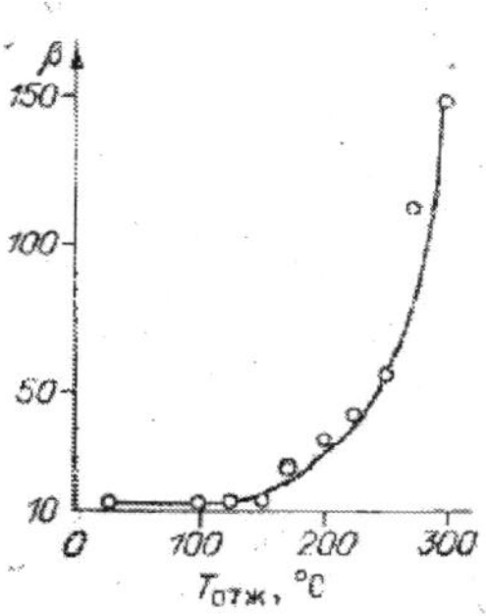

Figure 3.12. Isochronous annealing of silicon transistors irradiated with protons. T_{area}=250°C, Φ = $9\cdot10^{14}$ cm^2; proton energy 50 keV, annealing time 15 min at every temperature.

First experiment. Silicon transistors were irradiated with electrons with an energy of 3.5 MeV or protons with an energy of 50 keV at elevated temperatures. The results (Fig. 3.11 and 3.12) indicate that, during isochronous annealing, the restoration of the gain of transistors occurs at temperatures significantly lower than the irradiation temperature.

Second experiment. Silicon of p-type conductivity, containing oxygen at a concentration of $7\text{-}13\cdot10^{17}$ cm^{-3} with a specific resistance at room temperature of 7.5 Ohm·cm, was irradiated with electrons with an energy of 3.5 MeV at elevated temperatures. Then, during the process of isochronous annealing, the formation of a high-resistance near-surface layer was observed, which manifested itself in an increase in the sample resistance [66]. And again (Fig. 3.13) an increase in resistance was observed at temperatures below the irradiation temperature.

The third experiment. Let us consider the results of [67], the authors of which observed the kinetics of accumulation and annealing of A-centers at elevated irradiation temperatures. The irradiation temperature was selected at 375°C. At this temperature, the A-centers formed by irradiation at room temperature are almost completely annealed. The experiments were carried out according to the following schemes. Scheme 1: irradiation at a temperature of 375°C - turning off the beam and heater, cooling to room temperature, holding at room temperature, isothermal annealing at a temperature of 375°C. Scheme 2: irradiation at a temperature of 375°C - turning off the beam, isothermal annealing at a temperature of 375°C, turning off the heater and cooling to room temperature.

Table 3.1. A-center annealing kinetics

T_{area}, °C	Delay time, sec	A-center concentration, 10^{-14} cm^{-3}
20	0	1,27
375	0	1,10
375	180	1,11
375	360	0,75
375	1680	0,10

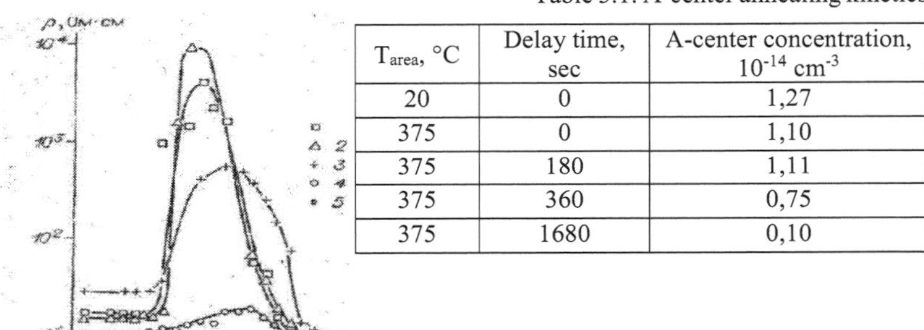

Fig. 3.13. Isochronous annealing of silicon irradiated with electrons. Silicon KEF-7.5, oxygen concentration (8-13)·$10^{17}$$cm^{-3}$, electron energy 3.5 MeV. Irradiation temperature, °C: 1 - 20; 2 - 300; 3- 350; 4 - 400; (1, 2, 4 - the dose of electrons 1,1·10^{16} cm^{-2}; 3 – 1,7·10^{16} cm^{-2}); 5 - unirradiated sample. Annealing for 15 minutes at each temperature.

From a comparison of the final concentrations of A-centers in experiments using two schemes, the authors of [67] expected to obtain information on the behavior of vacancies after switching off a beam of fast electrons. If the process of accumulation of vacancy - oxygen complexes is still far from saturation and the lifetimes of free vacancies are significant, then in the experiment according to Scheme 2, the sites of the formation of A-centers should be visible even after the beam is turned off or their annealing is delayed. The concentration of A-centers was determined from measurements of the temperature dependence of the Hall effect of the initial and irradiated samples, as well as at different stages of annealing. The results are shown in table 3.1. The main one is the delay in the annealing of the A-centers for the samples irradiated according to Scheme 2. Only at large delay times for turning off the heater does the annealing of the vacancy + oxygen complex begin to prevail.

The effect is clearly noticeable when compared with the results of isothermal annealing at a temperature of 375°C for samples irradiated according to Scheme 1. Over a period of 180 s, the concentration of the center under study decreased by 30%. The samples irradiated according to Scheme 2 underwent the same heat treatment, only immediately after the beam was turned off, the concentration of *A*-centers remained unchanged (see Table 3.1). The authors of [67] conclude that the lifetime of free vacancies at a temperature of 375°C is at least 180 s.

Thus, the above experiments allow us to draw the following conclusions.

1. At elevated irradiation temperatures, there is a large concentration of simple defects due to an increase in their lifetime.
2. The concentration of defects whose annealing temperature is lower than the irradiation temperature is not vanishingly small, but is determined by the ratio of the processes of defect formation during the continuous generation of at least one of the components and the annealing of these defects. As can be seen from table 3.1, the quasi-equilibrium concentration of defects can be by no means small.
3. A large concentration of primary defects (vacancies) increases the likelihood of the formation of complex multi-vacancy complexes.
4. During cooling after the beam is turned off, the concentration of complexes can increase as a result of the formation of these defects by the free components of the Frenkel pair.

We do not touch upon such a trivial effect of temperature as an increase in the mobility of the components of any associations.

The role of irradiation intensity is manifested in two ways: first, the charge state of the defect, and, consequently, its mobility, can change (see, for example, [19]); secondly, due to a higher concentration of primary defects, the probability of the formation of a multicomponent defect increases. An example of this is that all multi-vacancy defects are fixed in samples irradiated with ions or neutrons.

In conclusion, let us consider in more detail the work [64], in which the author in a wide range of doses analyzes physical processes leading to the enlargement and accumulation of specific types of radiation defects (for definiteness of the vacancy type) up to the formation of an amorphous state.

Processes are divided into three groups:

I. Quasi-primary processes of formation and stabilization of disordered areas

II. Coagulation processes involving mobile point defects, which, interacting with each other, as well as with defects formed at earlier stages of irradiation, contribute to the creation of new stable defects and enlargement of the "old" ones and lead to the accumulation of large radiation defects on average uniformly in volume irradiated layer. The spatial separation of the simplest defects is ensured by the presence of an activation barrier for annihilation of vacancies and interstitial atoms [68, 69], as well as their high mobility.

   For the processes of this group, the nature and amount of chemical impurities, structural defects of the initial crystal, the presence of elastic stress fields and their nature, charge states of interacting defects, and other factors affecting the activation barriers and mobility of interacting particles can play an extremely important role. The idea of the coagulation mechanism was put forward in [70, 71].

III. Heterogeneous processes of the formation of an amorphous layer at the surface of the crystal, as well as at grain boundaries and other major violations of the crystal structure.

The main factor determining the conditions for the spatial separation of vacancies and interstitial atoms in the case of heterogeneous defect formation is, most likely, relaxation of the fields of internal stresses associated with the interfaces.

In all cases, the efficiency of the mechanisms of the spatial separation of vacancies and interstitial atoms will be greatest at the initial stages of irradiation, when the crystal is still sufficiently perfect. As accumulation and enlargement of complex defects themselves become centers of annihilation. This leads to an equilibrium state between the processes of defect formation and their departure to previously formed defects i.e. with an increase in the radiation dose, the concentration of new defects does not actually increase. Under these "quasi-stationary" conditions, only sequential rearrangements of previously formed defects and the number of defects (in terms of point defects) accumulated at the previous irradiation stages are carried out. With a sufficiently long exposure to quasi-equilibrium, a certain "limiting" concentration of complex defects is achieved.

In [64], the entire dose range is divided into two sections: 1) the region of small doses ($0 < \Phi < \Phi^*$); 2) the region of large doses $\Phi \geq \Phi^*$. Here Φ^* is identified with a dose of amorphization.

In the low-dose range, the action of the spatial separation mechanism of the components of the Frenkel pairs will be taken into account by the parameter α_0 under the assumption that in the unit volume of the irradiated layer only mono-vacancies are generated with a speed of $\alpha_0 L^{-1}\varphi$, which does not depend on time in the considered dose range. Here φ is the particle flux density; L is the thickness of the irradiated layer. Parameter α_0 is numerically equal to the average number of "active" mono-vacancies introduced into the crystal by one primary particle. This parameter is not theoretically reliably calculated and is found experimentally.

The coagulation mechanism can be represented by a chain of interconnected quasi-chemical reactions of the form

$$\begin{aligned} &V + V \xrightarrow{\lambda_1} W_2 \\ &W_2 + V \xrightarrow{\lambda_2} W_3 \\ &\ldots\ldots\ldots\ldots\ldots\ldots \\ &W_j + V \xrightarrow{\lambda_j} W_{j+1} \end{aligned} \tag{3.10}$$

where V - movable monovacancies; W_j ($j \geq 2$) - fixed vacancy clusters consisting of ($j > 2$) monovacancies; ($\lambda_j \geq 1$) - reaction rate constants.

In the dose range, when the spatial separation of the simplest defects is weakened, the system of quasi-chemical reactions (3.10) must be supplemented by a chain of reactions of the following form

$$\begin{aligned} &W_2 + I \xrightarrow{\nu_2} V \\ &W_3 + I \xrightarrow{\nu_3} W_2 \\ &\ldots\ldots\ldots\ldots\ldots\ldots \\ &W_j + I \xrightarrow{\nu_j} W_{j-1} \quad (j > 2) \end{aligned} \tag{3.11}$$

where I mobile interstitial atom; ν_j ($j \geq 2$) rate constants of reactions of capture of interstitial atoms by vacancy clusters.

Further in [64], systems of equations are written that correspond to schemes (3.10) and (3.11) with the corresponding initial and boundary conditions, and their solution is given. We present only the final result to demonstrate the dose dependence of the concentration of vacancy clusters:

$$\tilde{W}_j(\Phi)/W_0 = \left(2^{j-z}/(j-1)!\right)(\Phi/\Phi_0)^{j-1}, \Phi/\Phi_0 \ll 1, \quad j \geq 2 \qquad (3.12)$$

$$W_j(\Phi)/W_0 = 2^{j-2}(\Phi/\Phi_0), \Phi/\Phi_0 \gg 1, \quad j \geq 2 \qquad (3.12b)$$

where W_0 and Φ_0 - theory parameters. From these formulas it follows that the concentration of any vacancy cluster containing j ($j \geq 2$) vacancies is a nonmonotonic function of Φ: in the intermediate dose range, the concentration reaches its maximum value, and further irradiation leads to its decrease [see formula (3.12b)].

The author [64] gives expressions of relative doses (Φ_{mj}/Φ_0) for maximum concentration (W_{mj}/W_o) clusters of j vacancies and half width maximum at the level 50% ($\beta_j^{(-)}$ and $\beta_j^{(+)}$). Calculation results for j = 2 - 6 are presented in table. 3.2. It can be seen that the dose at which the maximum concentration of a specific defect is reached W_j, exponentially increases with increasing j. In this case, the maximum concentration of this defect decreases rather slightly (according to law ~ $j^{-1/2}$), and the relative width of the maximum grows with j.

For definiteness, vacancy clusters were discussed. But experiments on electron microscopy have shown that upon ion irradiation [72, 73] or intense electron [74], large defects of the interstitial type are observed.

Table 3.2. Kinetics in the region of maximum defect concentration.

Defect (W_j)	Φ_{mj}/Φ_0	W_{mj}/W_0	$\beta_j^{(-)}$	$\beta_j^{(+)}$
Divacancy (j=2)	1,000	0,2500	0,2679	3,732
Trivacancy (j=3)	2,513	0,2036	0,2394	5,460
Tetravacancy (j=4)	5,009	0,1781	0,2546	5,695
Five vacancies (j=5)	9,161	0,1610	0,2507	6,137
Sixth vacancy (j=6)	16,070	0,1483	0,2389	6,693

A great influence on the kinetics of defect accumulation is exerted by the interaction of simple defects with various lattice imperfections. The following paragraph is devoted to this question.

3.4 Interaction of radiation defects with various imperfections of the crystal lattice

Thermal defects. It can be said that thermal defects are a well-studied defect. A large number of works are devoted to the study of this object in Germany [75–80]. Detailed literature can be found in [81]. Although the authors of Ref. [79] try to associate thermal acceptors in Germany with vacancy or divacancy + donor complexes, this conclusion, apparently, cannot be considered final. Only one thing can definitely be said: the presence of thermal acceptors leads to a decrease in the rate of introduction of radiation defects [80], that is, thermal defects in germanium are an effective sink for radiation defects.

Much attention was paid to the properties of thermal defects in silicon [81–83]. In a number of works [84–86], the interaction of radiation and thermal defects was studied. Irradiation with γ-rays during heat treatment [84] is equivalent to a decrease in the temperature of heat treatment. In the irradiated samples, a decrease in the concentration of thermal donors [84] and an increased yield of *A*-centers [86] were observed. The authors of [86] concluded that vacancies efficiently interact with thermal donors, which include from 2 to 4 oxygen atoms. They also proposed a process diagram explaining the increased yield of *A*-centers in the heat-treated material and the decrease in the concentration of thermal defects.

Thus, in both germanium and silicon, thermal defects are effective sinks for radiation defects and can play the role of coagulation centers for simple defects.

Dislocations. The deformation around the edge dislocation has an alternating character - there are areas of compression and tension. When a vacancy (interstitial atom) is absorbed, the extra half-plane shortens (lengthens) - the dislocation crawls. In Germany, crawling is only possible when heated to $T > 700°C$.

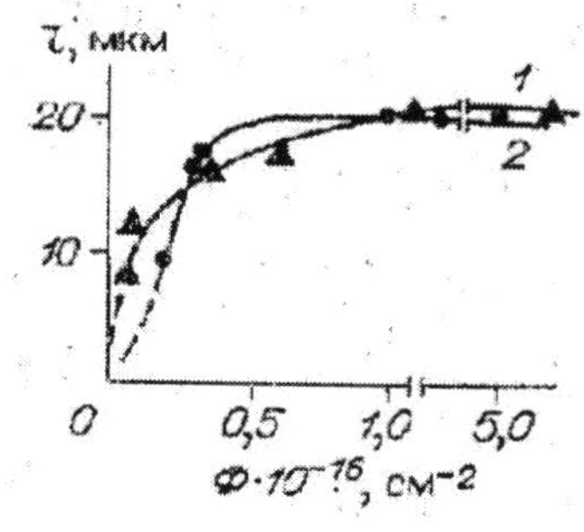

Figure 3.14. Dependence of the creep of the dislocation on the integral electron flux in *n*-type germanium (1) at an irradiation temperature of 430°C and *p*-type (2) at an irradiation temperature of 300°C.

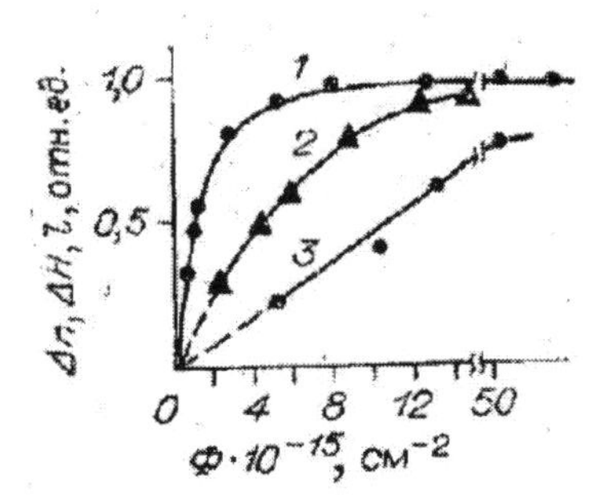

Figure 3.15 Dependence of the concentration of charge carriers (1), microhardness (2) and creep of dislocations (3) on the dose of radiation Φ.

Under electron irradiation, crawling is observed at substantially lower temperatures [87]. In the direction of creep, it was determined that interstitial atoms are being absorbed. The creep of dislocations was observed in the range of irradiation temperatures of 100–430 ° C [88]. Elevated irradiation temperatures for observing creeping, as noted by the authors of [88], are necessary for "loosening" impurity atmospheres around dislocations, that is, to facilitate the penetration of an interstitial atom to a dislocation core. The creep saturation (Fig. 3.14) is due to the interaction of newly formed interstitial atoms with accumulated vacancies and, accordingly, with a decrease in the influx of interstitial atoms to the dislocation core.

Another possibility of releasing dislocations from impurity atmospheres is quenching from relatively low temperatures [89]. Indeed, in quenched crystals, creeping was observed at room temperature [89].

Attention should be paid to one feature of the samples where dislocation creeping was observed — these crystals are enriched in vacancies (vacancy complexes). The mobility in such crystals is significantly lower than in samples without dislocations. The difference in mobility manifests itself even more after a small annealing (~80°C) [90]. The temperature dependence of mobility in crystals enriched in vacancies has a bell-shaped form with a sharp decrease in the mobility of charge carriers at low temperatures. According to the authors of [90], dislocations act as centers of condensation of radiation defects.

The results of a study of dislocation recombination radiation in n- and p-type germanium crystals also indicate an effective interaction of radiation defects with dislocations and a significant effect of impurity atmospheres on the characteristics of recombination radiation [91]. There are no reports in the literature about the creeping of dislocations in silicon. This is probably due to a higher concentration of impurities around dislocations than in germanium.

Dislocations as sinks and centers of condensation of radiation defects also affect the kinetics of accumulation of radiation defects. Since the various characteristics of a semiconductor material depend differently on the dose of the irradiating particles (Fig. 3.15), more information can be obtained by observing the dose dependence of several parameters of the crystal.

In [92], the effect of dislocations on the kinetics of accumulation of radiation defects was studied by observing a change in two characteristics: carrier concentration and microhardness. The first value is a characteristic of electrically active defects, the second gives an idea of all the defects accumulated in the crystal. Dose dependences of the rate of introduction of charge carriers and changes in microhardness are presented respectively in Fig. 3.16 and 3.17. The main results of this work are as follows.

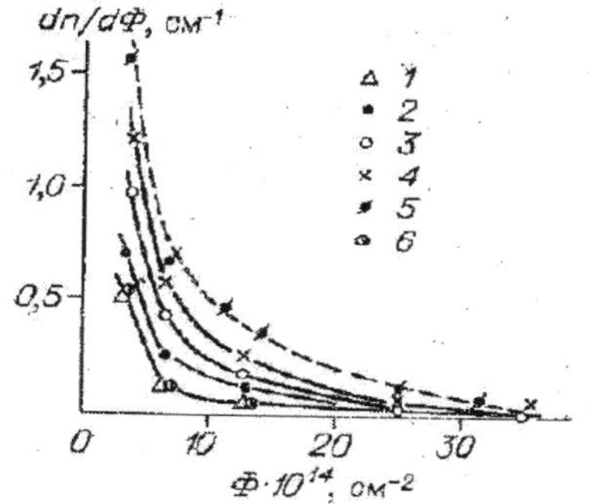

Figure 3.16. Dependence of the rate of change in the concentration of charge carriers on the radiation dose at various dislocation densities. N_d, cm^{-2}: 1-10^1; 2-10^3, 3-10^5; 4-10^7; 5-10^3, 10^5, 10^7 (match); 6-10^1 (5, 6- preliminary hardened from 200°C).

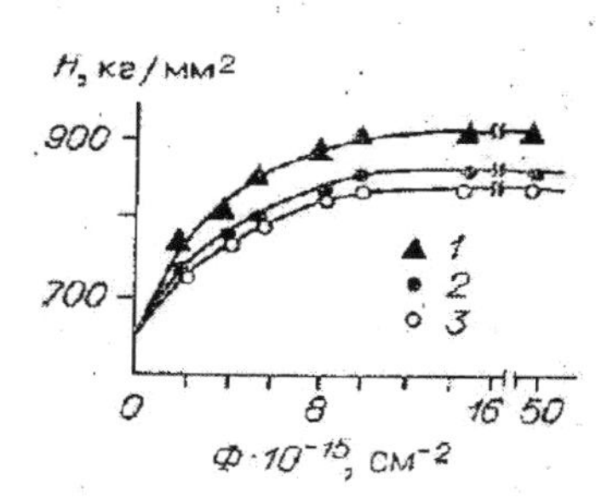

Figure. 3.17. Change in microhardness depending on the dose of radiation at different densities of dislocations.

1. The rate of introduction of electrically active radiation defects increases with increasing dislocation density at electron fluxes less than 4.10^{15} cm^{-2} (see Fig. 3.16).
2. A noticeable increase in microhardness H begins at electron fluxes greater than $2\cdot10^{15}$ cm^{-2}, saturation occurs at doses an order of magnitude greater than for the concentration of charge carriers, and the maximum increase in H depends on the density of dislocations (see Fig. 3.17). The rate of change in microhardness decreases with increasing dislocation density.
3. Quenching of samples before irradiation contributes to an increase in $dn/d\Phi$, and, very importantly, its dependence on the dislocation density disappears (dashed curve in Fig. 3.16). Annealing quenched samples before irradiation almost completely removes the hardening effect. Hardening does not affect the rate of change in microhardness.

4. Annealing of all samples irradiated with small doses (up to $5 \cdot 10^{15}$ cm^{-2}) completely restores the concentration of charge carriers and microhardness. After annealing at 300°C of samples irradiated with large doses (above $5 \cdot 10^{15}$ cm^{-2}), the concentration of charge carriers and microhardness in deformed crystals differ from the initial values (for $N_d = 10^5$ cm^{-2} $\Delta n/n_0 = 30\%$, $\Delta H/H_0 = 6\%$, for $N_d = 10^7$ cm^{-2} $\Delta n/n_0 = 50\%$, $\Delta H/H_0 = 9\%$, $\Phi = 5 \cdot 10^{16}$ cm^{-2}).
5. In a dislocation-free material, quenching before irradiation does not affect the rate of introduction of radiation defects. Heating in the deformation mode of a dislocation-free material does not change the value of $dn/d\Phi$. In a deformed sample of dislocation-free material, the rate of introduction of radiation defects increases sharply.

All these facts, according to the authors of [92], point to the large role of dislocations in the formation of radiation defects and do not fit into a simple model that takes into account the effect of dislocations only as sinks for interstitial atoms. According to this model, as a result of quenching, sinks for interstitial atoms become more efficient. In this case, an increase in $dn/d\Phi$ should be observed, and the dependence on the dislocation density should be preserved or even strengthened. It is clear that during irradiation there are some processes competing in capturing the simplest defects, most likely the interaction of the components of Frenkel pairs with dislocations and with chemical impurities, the concentration of which in the volume also changes with the introduction of dislocations and subsequent heat treatments.

The authors of this work suggest that interstitial atoms "stick" to some centers, and complexes including vacancies are considered responsible for changes in the concentration of charge carriers. A change in the concentration of "sticking" centers, which are played by both dislocations and chemical impurities (in particular, impurities that create atmospheres at dislocations), directly affects the number of interstitial atoms that can annihilate at a given temperature, and this determines the concentration of vacancies giving active centers. From this point of view, the rate of introduction of radiation defects depends on the concentration of "free" vacancies. Their concentration is the higher, the more interstitial atoms are spent on "sticking" centers, which serve as dislocations and chemical impurities. In deformed samples, the impurity is deposited on dislocations, freeing up the bulk, and then the main channel for the adherence of interstitial atoms is dislocations. Therefore, with increasing dislocation density, $dn/d\Phi$ also increases. Hardening performed before irradiation with thermal shock frees dislocations from impurity atmospheres, an impurity enriches the volume and becomes the dominant channel for the adhesion of interstitial atoms. The concentration of this impurity is the same for all samples of this ingot. It is difficult to say which impurity works effectively as a center of adhesion. One can confidently say only that it actively interacts with dislocations, decorating them.

The decrease in the rate of change in microhardness in samples with a high dislocation density is due to the fact that mainly point defects condense on dislocations, leaving fewer centers in the volume that impede the movement of dislocations arising under the indenter.

Interface between two phases. The role of the surface or interface between the two phases as a sink for radiation defects is especially pronounced during ion bombardment (see, for example, [93, 94]), and the effect is so great that it can even lead to amorphization of the surface layer. To illustrate, we give a description of the experiment from [93]. The study was conducted by transmission electron microscopy and fast electron diffraction by reflection of the $Si0_2$-Si system, previously irradiated with protons with an energy of 10 keV at a temperature of 200°C. An oxide film with a thickness of 500 Å was grown on the surface of *p*-type silicon using anodic oxidation and, after being bombarded with a proton flux $(1\text{-}50)\cdot 10^{15}$ cm^{-2}, was etched in hydrofluoric acid.

The choice of irradiation conditions was determined by the following: a) proton bombardment, creating a high density of simple defects, does not lead to amorphization; b) the proton energy is chosen so that the $Si0_2$–Si interface is in close proximity to the maximum energy loss of protons due to elastic collisions (~ 1000Å); c) the choice of the irradiation temperature was motivated by the need to reduce the role of conventional complexation channels and increase the role of defects distinguishable in an electron microscope.

A dose of $4\cdot 10^{16}$ cm^{-2} protons was sufficient to form a surface amorphous layer. It follows that, indeed, the interface is a sink for radiation defects, and their accumulation can lead to amorphization.

In [66], experiments are described that show that the accumulation of defects near the surface is also observed after electron irradiation and subsequent annealing.

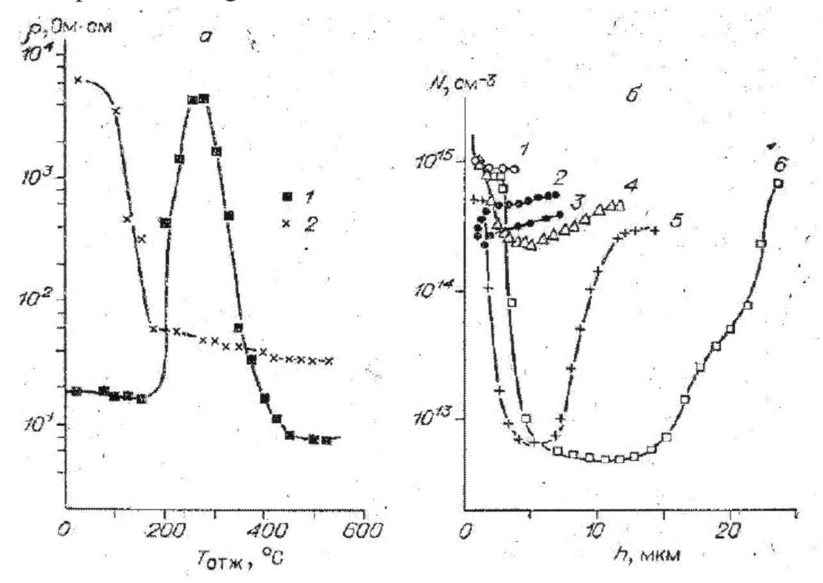

Figure 3.18. Change in the electrophysical characteristics of silicon during irradiation and annealing.

a - dependence of the specific resistivities of silicon on the annealing temperature: 1 - KEF-7.5; oxygen concentration $8\text{-}13\cdot10^{17}$ cm^{-3}, $\Phi=1\cdot10^{16}$ cm^{-2}; 2 - BKEF-20, oxygen-free, $\Phi=7{,}5\cdot10^{14}$ cm^{-2}; b - electron distribution in KEF-7.5 silicon after irradiation and annealing at T, °C: 1–500, 2–100, 3–200, 4–400, 5–300, 6–350. Annealing time 15 min, $\Phi=1{,}1\cdot10^{16}$ cm^{-2}, fast electron energy 3.5 MeV.

Silicon of *n*-type conductivity (KEF-7.5) was irradiated at room temperature with electrons with energies of 0.5-1.5 and 3.5 MeV. Isochronous annealing was carried out in the temperature range 100-600°C. The annealing results are presented in Fig. 3.18a. For comparison, irradiation and annealing of an oxygen-free sample (BKEF-20) were carried out. The oxygen concentration in the KEF-7.5 samples was $(8\text{-}13)\cdot10^{17}$ cm^{-3}, and in the oxygen-free (BKEF-20) less than $1\cdot10^{16}$ cm^{-3}. The experimental results are as follows: a) annealing of an oxygen-containing crystal in the range of 170 - 370°C leads to a “reverse” annealing - a sharp increase in the resistance of the sample; b) removal of the surface layer with a thickness of ~30 μm after annealing at 200–350°C sharply reduces the resistance of the sample, i.e., during the annealing, a high-resistance surface layer is formed; c) staining of the thin section with copper confirmed the presence of a high resistance layer near the surface. These results are confirmed by the results of measuring the capacitance of the Schottky barrier on irradiated and annealed silicon (see Fig. 3.18b); d) the thickness of the high-resistance layer increases with annealing time at a fixed temperature as $\sqrt{t}$, where t is the annealing time, and the activation energy of the formation of this layer is E ~0.8 eV; e) the high-resistance surface layer disappears after annealing at T≥400°C and is observed only in oxygen-containing silicon. It is assumed that the formation of the layer is associated with diffusion processes. The temperature region of its formation, which coincides with the region of decomposition of divacancies and *A*-centers, makes it possible to conclude that vacant components can be. The formation of a high-resistance layer is associated with the appearance of defects introducing E_c - 0.37 eV and E_c - 0.28 eV into the band gap. Complexes that compensate conductivity near the surface apparently include oxygen. But to indicate specifically the model or at least the full composition of the complex, experimental data are still lacking.

The interaction of simple defects with disordered regions was partially considered when discussing the properties of disordered regions. The interaction of disordered regions with defects is especially pronounced in the effect of radiation annealing [95, 96]. It is detected with combined irradiation. Thus, if large vacancy clusters (*VV*-centers [19]) are introduced into silicon by ion bombardment, then upon further irradiation with protons an increase in the number of *VV*-centers is observed, although proton irradiation itself does not introduce *VV*-centers [97].

The interaction of radiation defects arising from neutron irradiation with "biographical" defects is pronounced. This issue is discussed in detail in Chapter 4.

3.5 Annealing of radiation defects

In the narrow sense, annealing is understood to mean thermal irreversible dissociation of the defect in question. In the experiment, most techniques are selective. For example, a change in the charge state of a defect leads to the disappearance of the EPR spectrum in it. In the experiment, this is recorded as the disappearance, annealing of the defect. Annealing suggests that with increasing temperature and annealing time, the defect concentration should decrease. In experimental works, the term "annealing" is understood more broadly: this is a change (increase or decrease) in the concentration of defects. Therefore, sometimes for annealing, accompanied by an increase in the number of defects, the concept of “negative or reverse annealing” is introduced. Thus, annealing covers the following processes: 1) thermal decomposition of the defect; 2) the movement of the defect as a single entity to the stock; 3) the attachment of one of the components of the Frenkel pair to an existing defect (enlargement of the defect or annihilation at the center); 4) the separation of one of the components of the defect and its irreversible departure to the drain. As a rule, experimenters deal with defects that are reasonably stable at the measurement temperature and carry out the measurements themselves in minutes, or even hours and days after the end of irradiation. However, for the operating devices and circuits, which are sometimes very fast, the kinetics of the transition of the simplest primary defects to stable complexes is important. In a number of works [98-101], the kinetics of the formation of stable defects was studied on instrument structures after exposure to a short pulse of electrons or neutrons.

Although primary defects are thermally unstable at room temperature, the formation of defect complexes with impurities and with each other requires a finite time. For example, it was shown in [100] that the interaction of a silicon vacancy and interstitial oxygen with the formation of stable *A*-centers at 300 K in *p*-type silicon requires a time of the order of 10^{-2} sec. In the *p*-type material, this process proceeds much faster due to the lower activation energy of the negatively charged vacancy and ends in about 10^{-7} sec.

The effects of unsteady annealing (that is, annealing that occurs after the beam of irradiating particles is turned off) upon electron irradiation is masked by the presence of ionization [100]. After a decrease in ionization, only stable defects are observed.

Of particular interest due to the lack of powerful ionization is the study of annealing after irradiation of silicon with a fast neutron pulse [98-101]. The neutron pulse duration is 50-100 μs. The experiments were conducted on solar cells and transistors.

The effect was expressed by the annealing coefficient, which is the ratio of the number of defects at some point in time t to the concentration of stable defects, i.e. $k=N(t)/N_{stab}$. By definition, this coefficient is greater than or equal to one. In fig. 3.19 shows the dependence of the annealing coefficient on time for transistors irradiated at different temperatures. A strong temperature dependence is clearly visible: annealing coefficient increases with decreasing temperature.

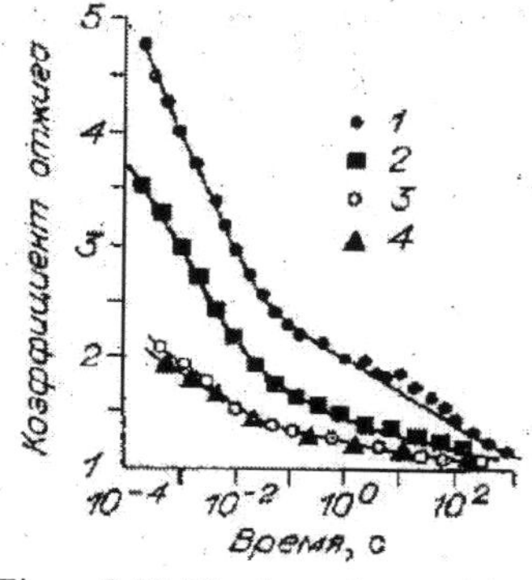

Figure 3.19. The dependence of the annealing coefficient on the time of the 2N914 transistor at various temperatures [98]. I_{av}=200 μA, I_{max}=2 mA, repetition rate 10 kHz, pulse duration 10 μs; Tann., K: *1*-213, *2* -268, *3*- 300, *4* - 348.

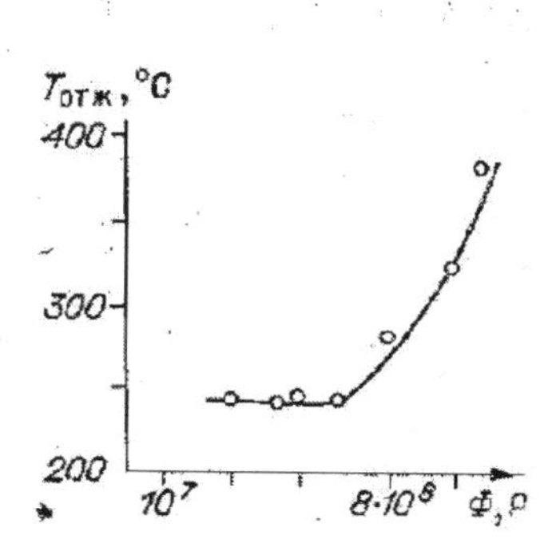

Figure 3.20 Dose dependence of the annealing temperature of *A*-centers [102].

From experiments on solar panels, it was possible to determine that in *p*-type silicon, processes occur with an activation energy of 0.3 eV, and in the *n*-type, 0.1 eV. The study of the dependence of the annealing characteristics in transistors on the injection level showed that increasing it accelerates annealing.

After 1000 s after the end of the pulse, we can assume that only defects stable at room temperature remained. The annealing of the simplest complexes of divacancies, *E*- and *A*-centers, more precisely, about their temperature stability, has already been mentioned (see section 3.2 of this chapter). Here we point out some factors that can change the annealing temperature of a particular complex. The most important is the radiation dose. It was shown in [102] that an increase in the dose of gamma rays increases the annealing temperature of *A*-centers (Fig. 3.20). The authors do not explain what caused the dependence of the annealing temperature of the *A*-centers on the dose of γ quantum, but as a possible explanation they give the following hypothesis: the *A*-center formed near a defect, for example, a dislocation, is annealed at low temperatures, and that A the center that arose in the undisturbed region of the crystal has a significantly higher annealing temperature. At low doses, *A*-centers are formed mainly near large defects and have a low annealing temperature. With an increase in the dose in the total concentration of *A*-centers, an increasing proportion falls on associations formed in the undisturbed region of the crystal, i.e., *A*-centers with a high annealing temperature.

This hypothesis was confirmed in [103], where it was shown that the A centers formed near the crystal surface are annealed at 200–220°C. In silicon containing a low oxygen concentration, an increase in the temperature of annealing of divacancies is observed with an increase in the dose of bombarding particles. But the explanation is different. The delay in annealing is explained by the simultaneous action of two processes — annealing (decay) of divacancies and the formation of new divacancies when two vacancies are combined. The second process makes a noticeable contribution only if there is an additional supply of volume to vacancies, for example, due to the decay of some other vacancy complexes. The effectiveness of "feeding" vacancies in the crystal volume by an electron beam and an increase in the stability region of *A*-centers has already been mentioned (see section 3.3 of this chapter and [67]).

Dislocations have a significant effect on the kinetics of annealing [104]. In germanium, the annealing rate of defects responsible for changes in the concentration of charge carriers decreases with increasing dislocation density. In annealed samples, upon annealing, a sharp decrease and anomalous behavior of the temperature dependence of charge carrier mobility are observed. During the annealing process, as a result of the decomposition of the complexes, mobile point defects arise and the competition laws governing the capture of mobile defects by dislocations and stable defects come into force. This contributes to the emergence of "clouds" of defects around dislocations, rearrangements of defects, which will result in either annihilation or the formation of more stable defect associations.

Very eloquently evidence of rearrangements of defects during isochronous annealing, the data of [25, 53], presented in Fig. 3.21. In silicon irradiated with neutrons, the disappearance of some defects causes the appearance and increase in the concentration of others.

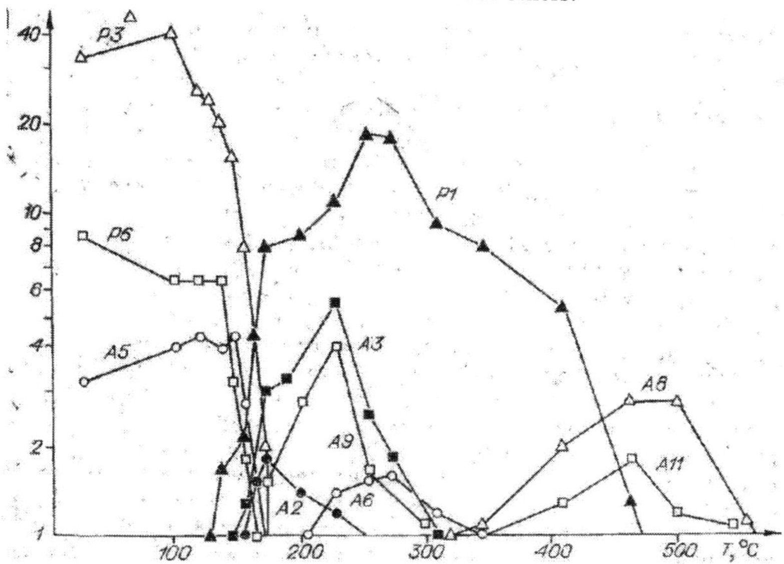

Figure 3.21. Change in the relative number of paramagnetic centers during annealing.

Annealing of large defects was studied in samples irradiated with ions and neutrons. The effects under these types of irradiation were similar (when calculating the radiation dose and the distribution of defects). In connection with the wider use of ion doping, there is more actual material specifically for the annealing of radiation defects in layers irradiated with ions. Moreover, much attention is paid to those annealing modes in which the electrical activity of the dopant is maximally manifested. More detailed information on radiation defects in ion-doped layers, irradiation and annealing regimes can be found in monographs [19, 20, 105-107]. Here, we briefly discuss the annealing and rearrangement of defects in silicon layers irradiated with ions (the annealing of silicon irradiated with neutrons is considered in Chapter 4).

Most of the large lattice defects arising from the incorporation of ions into silicon and germanium are annealed at temperatures of 600 ° C and higher. Measurements of the channeling effect show that there are distinct stages of annealing. It was established by electron microscopy that dislocation loops are formed upon annealing of disordered regions mainly at temperatures above 600 ° C. In silicon and germanium after irradiation with heavy ions, the character of annealing is highly dose-dependent. In the low-dose region, when the particle tracks are separated from each other, the crystal is restored at 180 ° C in germanium and at 260 ° C in silicon (Fig. 3.22). At high doses, when an amorphous layer is formed, recrystallization occurs at 380 ° C in Germany and at 570 ° C in silicon. These temperatures correspond to the medians of the 10-minute isochronous annealing curves. Changing the annealing time shifts the indicated temperatures. It is also obvious that in the annealing characteristics there should be some transition from the case of separate spatially separated disordered regions to the case of a continuous amorphous layer. In this transitional dose range, the annealing kinetics depends on how much of the disordered zones surrounding each track overlap. An example of the annealing characteristic for such a transitional dose is an intermediate curve for a germanium sample (see Fig. 3.22).

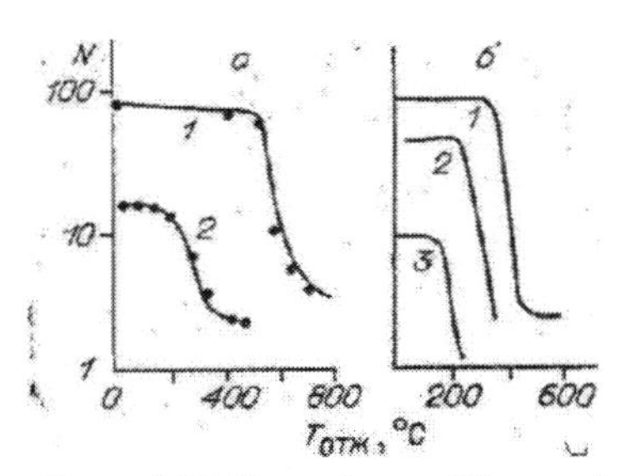

Figure 3.22. Dependence of the number of defects on the annealing temperature of (a) silicon and (b) germanium [105]. a - bombardment by antimony, arsenic and gallium ions (*1* - large dose, *2* - small); b - the same with indium ions (*1* - large dose, *2* - intermediate, *3* - small).

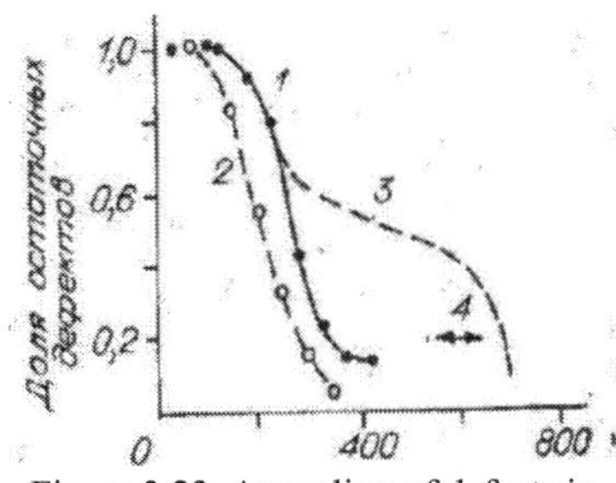

Figure 3.23. Annealing of defects in silicon irradiated by fast neutrons and subjected to ion bombardment. *1*-ion-doped samples (40 Kev, antimony, $1 \cdot 10^{13}$ cm^{-2}); silicon irradiated with fast neutrons: *2* - $\Phi = 4 \cdot 10^{16}$ cm^{-2}, *3* - $\Phi = 5 \cdot 10^{19}$ cm^{-2}, 4 - crystallization region of the amorphous layer.

In the case of isolated disordered regions in samples subjected to ionic implantation, the course of annealing of divacancies and the total number of defects repeats the course of annealing of divacancies and changes in the intensity of transmitted X-rays in samples irradiated with fast neutrons (Fig. 3.23). This shows that not only the nature of the disordering is the same in these two cases, but also the annealing behavior is due to the migration and dissociation of defects of the same type. The nature of isochronous annealing in crystals containing only point defects and in samples with disordered regions is different. This effect is illustrated in fig. 3.24, which shows the characteristics of annealing of vacancy – oxygen pairs obtained by measuring optical absorption. In a material irradiated with electrons, the intensity of the absorption line decreases, which indicates the dissociation of vacancy – oxygen pairs. In the material irradiated with neutrons, the absorption initially increases and only then decreases. Such an annealing process is explained by vacancy liberation from clusters and their capture by impurities.

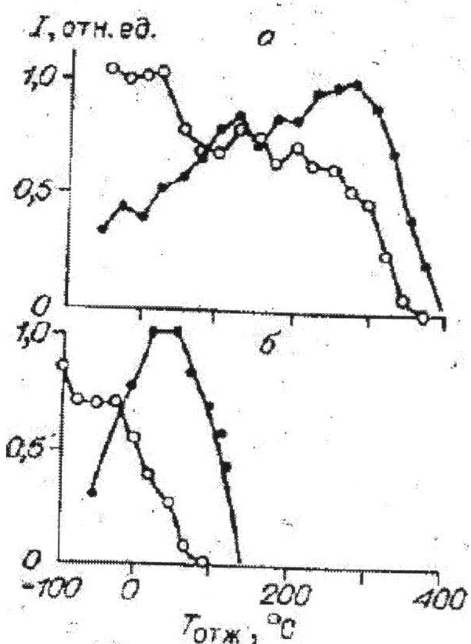

Figure 3.24. Dependence of the intensity of the infrared absorption line associated with the vacancy – oxygen complex on the annealing temperature [105].
a - *n*-type silicon; *b* - *n*-type germanium. Filled points - neutron irradiation, light - electrons

In the study of the restoration of the electrophysical characteristics of the disturbed layer in silicon, the stages of annealing below 600 ° C and above this temperature are considered.

T<600°C. As already mentioned, in silicon at a low dose of doping there is an annealing step at a temperature of 300°C. When analyzing the electrical characteristics of ion-doped layers, it was found that an inversion of the conductivity type usually requires annealing at 300-350°C. At T>350°C, the concentration of carriers in the ion-doped layers increases. This is due to the fact that the defects are in the immediate vicinity of the impurity and their manifestation from the impurity is necessary for the manifestation of its electrical properties. In this, the picture differs significantly from the situation when irradiated with fast neutrons, where the clusters are not connected with the dopant. At an annealing temperature of 400°C or more, the presence of clusters is manifested in a small mobility, the presence of compensating centers, and a strong temperature dependence of the concentration of charge carriers in ion-doped layers. Electrical measurements give an integral characteristic of annealing. Additional information can be obtained by observing the annealing of a specific center. In [19], data were presented on the annealing of a vacancy cluster — the *VV* center. The results are as follows.

1. With increasing dose and energy of the bombarding ions, the temperature of the complete annealing of the *VV*-centers increases (see Fig. 3.24). This is observed after irradiation with ions of both inert gases and silicon.
2. At least two stages can be distinguished on isochronous annealing curves: the first corresponds to a temperature of 150-200°C, the second - approximately 600°C. Moreover, the second stage is observed when silicon is irradiated with high-energy ions and at doses that are large in doses of amorphization.

T~600°C. If irradiation is carried out at room temperature (or lower) with doses sufficient for the formation of an amorphous layer, the concentration of electrically active impurities generally increases at an annealing temperature of ~ 600°C. At lower doses, the annealing curves show that the concentration of electrically active impurities reaches a maximum only at annealing temperatures ≥ 800°C. The same situation is observed with nuclear doping of silicon: annealing temperatures of ~ 800°C are required for the manifestation of the electrical activity of phosphorus in silicon.

1000°C > T > 600°C. In this temperature range, the maximum electrical activity of the impurity (T> 700 - 750°C) appears, when the dose of which was less than the dose of amorphization; deep and compensating centers are annealed. The electrophysical characteristics of ion-doped layers are close to those of bulk material.

It must be emphasized that, although the presence of different stages of annealing is pronounced, nevertheless, after annealing, a noticeable number of lattice defects can remain in the crystal. Electron microscopy studies confirmed this [108]. Upon annealing above 600°C, dislocation loops lying in the (111) plane, edge dislocations, and also dipoles parallel to the <110> direction appear. With increasing annealing temperature, dislocation loops increase from about 100 Å in diameter at 700°C to a maximum size of about 1000 Å at 800°C. Their size does not change when the temperature rises to 900 ° C. The discovered dislocation loops consist of interstitial atoms [108-111].

T>1000°C. Systematic studies of the defective structure in this region of annealing temperatures were not carried out. There are only a few studies on the layers annealed at such high temperatures. In [112], electron layers were used to study silicon layers doped with boron and phosphorus ions. In it, the authors note that at low doses of phosphorus (3-50 μC/cm^2), structural defects are not detected. At doses of 50-800 μC/cm^2, Frank loops are observed, the size of which increases from 200 - 300 Å to several microns with increasing radiation dose, temperature and annealing time. At high doses (>> 800 μC/cm^2) of phosphorus, dislocation networks of mismatch are observed. The same nets are found when large doses of boron are introduced.

A study of the kinetics of network formation yielded the following results. When heated to 1050°C, the formation of a large number of dislocation loops was noted. If the heating was carried out to 1100°C, then rod-shaped defects and dislocation loops of the interstitial type appeared in the layer. An increase in temperature to 1150°C led to the complete disappearance of dislocation loops and the formation of an ordered network.

In most annealing works, only experimental results are presented. Only in Refs [113, 114], the obtained data are conceptually interpreted and the theory of annealing of complex defects is proposed. By annealing, the author [114] understands not only the decay or enlargement of a defect, but also a change in the electronic structure, ie, the absence of a substance exchange between the defect and the matrix. It conditionally divides defects into two categories - small and large. If vacancies are part of small (decaying) defects, then mono-vacancies will be the dominant mobile defects. Each vacancy captured by a vacancy cluster leads to its enlargement. Some vacancies are lost at sinks - defects of the interstitial type, dislocations, etc. The author calls this situation annealing according to a scheme with external annihilation.

When interstitial atoms, which are the dominant carriers, enter into the composition of small defects, grinding of large vacancy clusters is observed. Such processes are considered as annealing according to the scheme with internal annihilation.

As applied to vacancy clusters, the annealing scheme with external annihilation has the form:

$$
\begin{gathered}
W_j \rightleftarrows W_{j-1} + V \\
\dots\dots\dots\dots\dots\dots \\
W_2 \rightleftarrows V + V \\
V \to S_\upsilon, \quad (j > 2)
\end{gathered}
\tag{3.13}
$$

where V - monovacancy; W_j(j ≥2) - vacancy cluster containing j vacancy; S_υ - drains for mono vacancies, due to which there is an irreversible decrease in the number of vacancies in the defective region.

A defect is considered “small” if it contains less than Q vacancies. For minor defects, only the reaction of thermal dissociation takes place.

$$
W_j \to W_{j-1} + V, \quad 2 \le j \le Q-1 \tag{3.14}
$$

and for large - only a bimolecular reaction

$$
W_k + V \to W_{k+1}, \quad k \ge Q \tag{3.15}
$$

Based on the solution of the corresponding system of differential equations, the author [114] concludes that small defects with a sufficient annealing time completely disappear. The concentration of large defects with $t \to \infty$ tends to nonzero asymptotic values. The annealing rate is highly dependent on time.

If among the initial large defects there is one or several types of defects, the concentration of which is significantly higher than the concentration of the others, then for the latter there can be reverse annealing, i.e., a relative increase in the concentration during the final time interval at the intermediate stages of annealing.

For vacancy clusters, annealing according to a scheme with internal annihilation is represented as

$$
\begin{gathered}
S_I \to I \\
W_1 + I \to 0 \\
W_2 + I \to W_1 \\
\dots\dots\dots\dots\dots \\
W_j + I \to W_{j-1}, j > 2
\end{gathered}
\tag{3.16}
$$

where I - own interstitial atom; S_I - "Source" of interstitial atoms.

In [114], all versions of the above scheme are analyzed in detail. The kinetics of annealing of vacancy clusters according to this scheme is determined by the processes accompanying the decay of interstitial defects. Due to insufficient knowledge of such defects, only the most general considerations about the nature of these processes can be expressed at present.

Let us dwell on one more procedure, which, according to the results, is similar to thermal annealing, and to a decrease in the concentration of existing defects during additional irradiation — by radiation annealing. The main reasons for this annealing are as follows.

1. The interaction of intrinsic point defects with large defective complexes, which leads to the decay of the latter or their rearrangements.
2. Radiation ionization, under the influence of which, as a result of a change in the binding energy and charge states, defective associations decay or energy thresholds change for annihilation of particles.
3. Radiation-stimulated interaction of impurity atoms (present or introduced by radiation) with defective complexes, leading either to their decay or to rearrangements with loss of activity in one or more parameters (electrical, optical, magnetic, etc.).

Radiation annealing as the interaction of point defects with large complexes was described in [115, 116]. To generate point defects, we used irradiation with electrons [115, 116] or protons [115]. In [115], an attempt was made to accelerate the annealing of disordered regions introduced by ion bombardment (argon, 40 keV) by preliminary irradiation of samples with fast electrons.

It turned out, however, that irradiation at room temperature up to doses of 10^{17} cm^{-3} did not noticeably affect subsequent annealing. Irradiation at T=250°C is more effective. A noticeable increase in the annealing rate was observed in the case when the dose of argon ions was small and no overlap of the displacement peaks occurred. It has been suggested that electron irradiation may be useful in annealing individual disordered regions created by neutron irradiation.

This assumption was experimentally verified and confirmed in [116]. It studied the effect of electron bombardment on the properties of silicon previously irradiated with fast reactor neutrons at a temperature of no higher than 70°C. The electron energy was 10 MeV, the dose was $4 \cdot 10^{18}$ cm^{-2}, and the irradiation temperature did not exceed 25°C. Then, isochronous annealing was carried out in the temperature range 100–800°C and the dependence of the IR absorption spectra in the wavelength range of 2–25 μm on the annealing temperature was studied. It was found that electron irradiation strongly attenuates absorption in the range of 4.5–14.5 μm, which usually appears in neutron-irradiated samples after annealing at a temperature of 500°C.

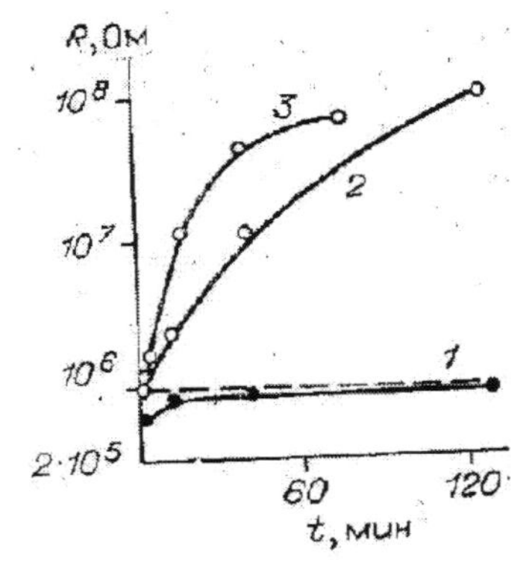

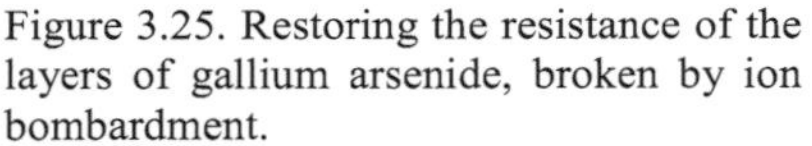

Figure 3.25. Restoring the resistance of the layers of gallium arsenide, broken by ion bombardment.
T_{ann} °C: *1*-250; *2*, *3*-250 with simultaneous irradiation with protons, energy 10 keV, current density 0.5 μA/cm^2: *2* - 0.15, *3* - 0.3.

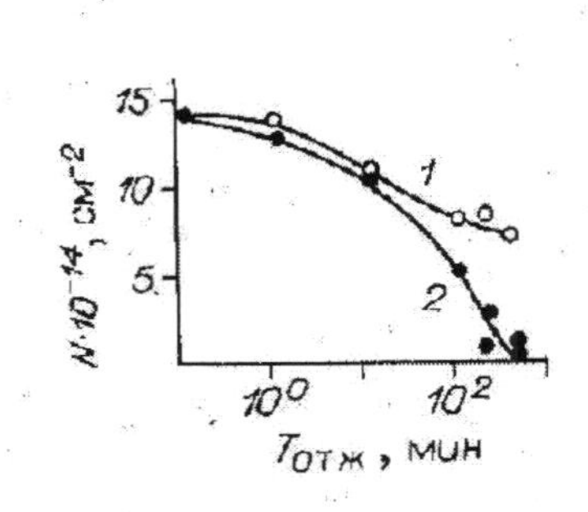

Figure 3.26. Annealing of paramagnetic centers in silicon amorphized by ion bombardment.
T_{ann}, °C: *1*-200; *2*-200 with simultaneous irradiation with protons, energy 10 keV, current density 0.5 μA/cm^2.

The role of protons as an annealing agent is shown in Fig. 3.25 and 3.26. In the latter case (Fig. 3.26), the authors of [117] emphasize that the "annealing" of paramagnetic centers is accelerated due to their interaction with hydrogen atoms. This interaction occurs when the hydrogen concentration is comparable to the concentration of paramagnetic centers. It was noted in [117, 118] that if the interaction of a defect with an impurity atom leads to the disappearance of the observed properties (for example, paramagnetism), then these properties can be restored by additional irradiation.

Recently, a new annealing technique, laser annealing, has been increasingly used. Based on the possibility of accelerating the annealing of defects during strong ionization and changing the charge state of defects, laser radiation was used to anneal ion-doped layers. Acceleration of annealing was already recorded in the first experiments combining heating and laser radiation [119]. When using powerful pulsed radiation, it was possible to completely abandon the usual thermal annealing, and sometimes to obtain qualitatively new effects [120-122]. In this case, although the role of heating cannot be excluded, the annealing process itself, undoubtedly, should be considered radiation-stimulated. Table 3.3 illustrates the effectiveness of laser annealing. Laser annealing of silicon irradiated with neutrons has not yet been described in the literature. Our experiments, conducted with the active participation of E.V. Nidaev, allow us to draw a preliminary conclusion that in neutron-irradiated silicon, laser pulsed annealing is in many respects equivalent to thermal annealing.

Table 3.3. The Hall method of measuring the electrophysical parameters of ion-doped layers on p-type silicon, obtained using laser and thermal annealing [122]

Bombardment mode			Laser annealing			Thermal annealing		
Ion type	Energy, keV	Dose, ions/ cm^2	$N_s \cdot 10^{-15}$ cm^{-2}	μ, cm^2/ /(V·s)	$\sigma_s \cdot 10^{-3}$, Ohm^{-1}	$N_s \cdot 10^{-15}$ cm^{-2}	μ, cm^2/ /(V·s)	$\sigma_s \cdot 10^{-3}$, Ohm^{-1}
P^+	40	3.10^{15}	1,5	66	16,5	1,2	68	2,3
P+	80	6.10^{15}	5,0	45	36,0	2,5	60	24,0
P+	150	2.10^{16}	9,0	40	57,0	5,2	58	48,0

Note. Laser annealing mode: λ=1,06 μm, τ=20 nsec, W=2,5·10^7 W/cm^2. Thermal annealing mode: T_{ann}= 800°C, t_{ann}=30 min.

Backscattering experiments (Fig. 3.27) showed a decrease in the defectiveness of ion-doped layers in the case of penetration at elevated temperatures due, apparently, to radiation-stimulated diffusion. We will dwell on this phenomenon in the next section.

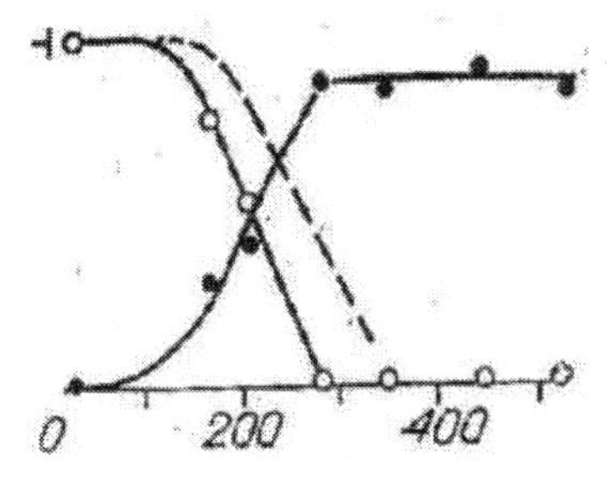

Figure 3.27. Influence of the interstitial temperature on the substitute component of antimony (filled circles) and on the number of point defects in silicon (light circles).
The dashed line for comparison shows the decrease during annealing of the number of defects in the layers doped with antimony at room temperature.

3.6 Radiation-accelerated diffusion

When considering radiation-controlled (accelerated or delayed) diffusion in semiconductors, two directions were identified: 1) diffusion, controlled by excessive structural defects; 2) diffusion accelerated by ionization.

A typical technique for observing radiation-accelerated diffusion in both silicon [123] and germanium [124] is to detect the displacement of the *p- n-* junction upon irradiation of a part of the sample with electrons or protons. The main results of these first works are as follows.

1. The diffusion shift of the transition takes place when exposed to protons with an average mean free path both comparable to and less than the depth of the transition.

2. The diffusion acceleration effect is insensitive to the charge state and the ionic radius of the impurity.

3. The acceleration of diffusion is observed at temperatures not lower than 800°C for silicon and 400°C for germanium.

4. In the temperature range 900–1200°C (for silicon) and 600–950°C (for germanium), the diffusion acceleration does not depend either on the particle flux density or on the irradiation temperature.

5. The temperature dependence of the diffusion rate, which appears at irradiation temperatures below 900°C for silicon and 600°C for germanium, shifts to lower temperatures with increasing radiation intensity.

The specified direction was further developed in [125-127]. The monograph [128] provides a detailed review of works up to 1975 on radiation-accelerated diffusion. In most studies, the effect is explained by an increase in the diffusion coefficient due to the appearance of excess vacancies upon irradiation. An alternative is the diffusion of atoms at interstitial positions. The role of irradiation in this case is reduced to the transfer of atoms from the site of the crystal lattice to the interstitial position. However, these two mechanisms do not allow explaining the totality of the experimental data. In [128], other possible mechanisms for accelerating diffusion are described; emphasis is placed on ionization mechanisms. But experimentally, it is very difficult to single out the role of ionization in accelerating diffusion, since several factors act simultaneously. One of such factors is the effect of entrainment by the "photon wind" [129].

In most experiments, the displacement of the p-n junction did not exceed 1 μm; only in [130, 131] did the impurity penetrate several microns. [130] used *p*-type silicon samples (KDB-10) doped with phosphorus ions with energy 100 KeV, doses 2.10^{15} and 5.10^{15} cm^{-2}. Direction of exposure <111>.

Samples with embedded phosphorus ($5 \cdot 10^{15}$ cm^{-2}) were irradiated with hydrogen ions at a Van de Graaff accelerator at a temperature of 900°C. The current density is 4 μA/cm². The concentration profiles of charge carriers were recorded by measuring the surface resistance by the van der Pauw method with successive removal of silicon layers by anodic oxidation. To obtain the initial impurity distribution profile, we used a sample annealed at 900°C. The annealing time and the irradiation time at the accelerator are 1 hour. The results are presented in fig. 3.28. The position of the shelf on curve 3 coincides with the projection of the ion path H_2^+, energy 500 KeV. When irradiated with hydrogen ions with an energy of 1 meV, the mean free path is ~6 μm and lies much deeper than the leading edge of the concentration profile.

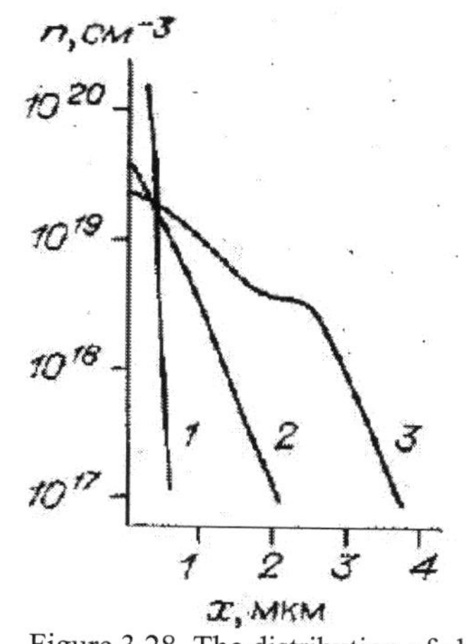

Figure 3.28. The distribution of charge carriers in depth.
1-annealing at a temperature of 900°C, 1 hour; *2* - irradiation with ions H_2^+, E=1 MeV; *3* - is the same, E=0.5 MeV; T_{irr}=900°C

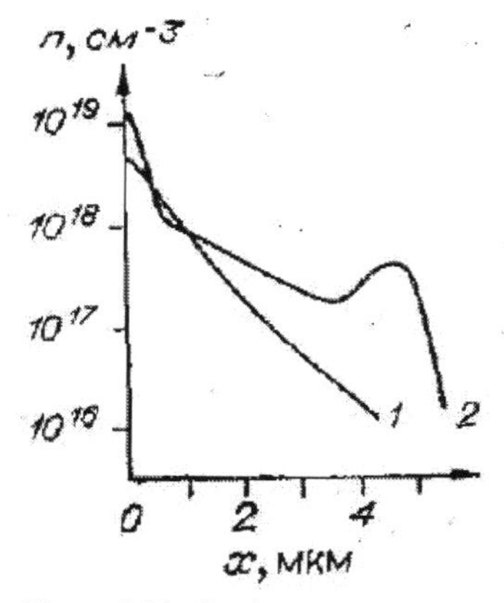

Figure 3.29. Carrier concentration profiles obtained by irradiation with H_2^+ ions, E=1 MeV, T_{irr}=900°C (*1* - j= 4 μA/cm*2*, 1 hour; *2* - j=0.5 μA/cm², 8 hour).

The acceleration of diffusion is naturally explained by the interaction of the impurity with radiation defects introduced into the crystal by irradiation. But with what? Either with those that came from the region of maximum elastic collisions, or with those that were generated near the impurity front.

The diffusion length of the radiation defects responsible for accelerating the diffusion of the impurity is apparently small, since otherwise there should be lateral diffusion of the impurity. However, it was shown in [132] that lateral diffusion is not observed upon proton irradiation, and the diffusion length of defects is much less than 1 μm. In addition, the large diffusion length of defects indicates their weak interaction with lattice imperfections, including impurities. The authors of [130] suggest that the impurity is accelerated by the generated radiation defects without their long-range diffusion. The mechanism of such diffusion is not specified. In Fig. 3.29 it is shown that for the formation of an impurity profile, the irradiation time, and not the dose, is important.

The work [131] describes the acceleration of phosphorus diffusion in germanium. The experiment used germanium washers doped with phosphorus ions with an energy of 100 keV (projected range of phosphorus ions ~ 0.08 μm) and a dose of 3.10^{15} cm-2 at room temperature. Irradiation with hydrogen ions of H_2^+ with an energy of 600 keV was carried out at temperatures of 400, 500, and 600 ° C and with an energy of 1000 keV at 600 ° C using a Van de Graaff accelerator. In all experiments, the ion current density was 1 μA / cm2, and the irradiation time was 2 h. The surface conductivity and effective carrier mobility were measured. From these data, the electron concentration distribution profile was calculated, which was identified with the phosphorus distribution.

The results were as follows:

1. After irradiation with hydrogen ions, the coefficient of utilization of the embedded phosphorus impurity is small, especially after irradiation at temperatures of 500 and 600°C (the coefficient of utilization is 0.37 after irradiation at 400°C; 0.008 after irradiation at 500°C; 0.018 after irradiation at 600°C; 0.72 after thermal annealing at 500°C for 2 hours).

2. After 400-degree irradiation, two peaks of increased impurity concentration are observed: the first is near the surface, the second is near the projected path of hydrogen ions with an energy of 600 keV (Figs. 3.30 and 3.31).

3. After 500-degree irradiation, a region of p-type conductivity is observed (see Fig. 3.30) with a length of up to 22 μm.

4. After 600-degree irradiation, the p-type region is smaller, but the electron concentration in the enriched layer (peak) is much higher.

Irradiation with hydrogen ions of control samples of germanium specially not doped with phosphorus did not lead to the formation of a n-type conductivity layer.

5. An increase in the energy of hydrogen ions, ceteris paribus, leads to a larger extent of the p-type region of conductivity and to a decrease in the electron concentration in the impurity peak (see Fig. 3.31).

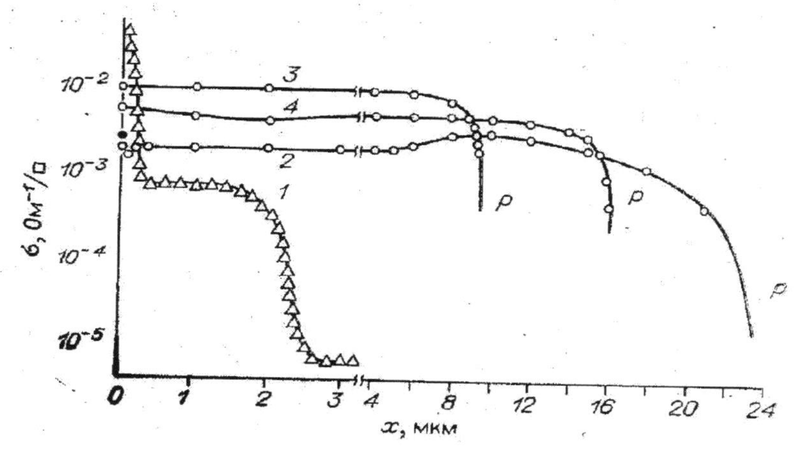

Figure 3.30. Normal distribution of surface conductivity in germanium. $T_{irr.}$°C: *1*-400, *2*-500, *3*, *4*-600. Energy H_2^+, keV: *1*-*3*-600, *4*-1000. For curves 2-4- it is indicated where it starts *p*-type conductivity.

The authors of [131], explaining the penetration of an impurity far beyond the projected range of irradiating ions, suggest that electrodiffusion of phosphorus ions takes place in a field created by unevenly distributed nonequilibrium excess charge carriers. Alternatively, they indicate the possibility of energy transfer to an impurity atom or a "phonon wind" [129], or elastic waves, which arise, for example, during the release of energy during the annihilation of simple radiation defects.

The above examples are sufficient to show that irradiation at elevated temperatures can significantly distort the expected distribution of the impurity. One can expect the manifestation of the effect of radiation-accelerated diffusion during radiation doping of silicon due to reactions on slow neutrons. This effect can especially manifest itself when a sharp boundary is needed between the doped and undoped parts of the crystal, if doping occurs under the influence of nuclear reactions. Moreover, the border can be blurred not only when irradiation is carried out at elevated temperatures. Acceleration of diffusion is also possible during annealing of defects [105, 106, 128]. Based on this possibility, one should choose the annealing mode of crystals, in which the sharpness of the boundary between the doped and undoped parts of the crystal is significant.

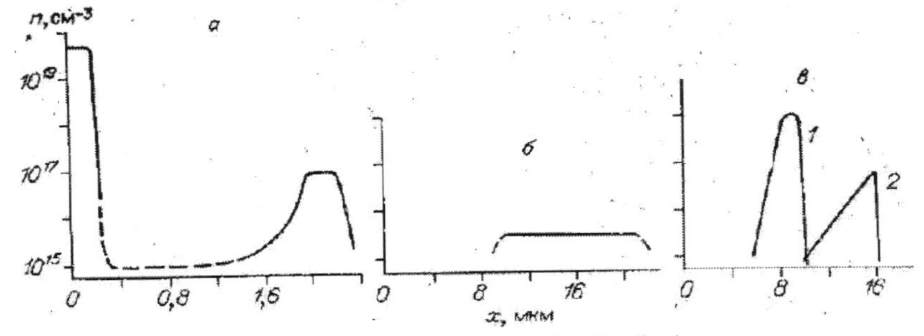

Figure 3.31. Electron concentration distribution.
a) T_{irr}=400°C, ion energy H_2^+ - 600 keV; b) T_{irr} = 500°C, energy H_2^+- 600 keV; c) T_{irr}=600°C, energy H_2^+, keV: *1* - 600, *2* - 1000.

It is clear from the foregoing that in silicon both the types of defects that have arisen and the processes occurring during irradiation and annealing are well studied. For other semiconductor materials, such specific data is much less.

3.7 Radiation-induced defects in germanium

Despite the fact that radiation effects in germanium have been studied for a very long time, concrete results on the identification of defects have been obtained much less than for silicon. Apparently, the main reason for this is that it is difficult to apply electron paramagnetic resonance to germanium. But the nature of most (if not all) defective associations (and simple defects) was deciphered using EPR. A large number of works (in the years 1950-60) are devoted to the study of defects arising from the low-temperature (~4.2 K) irradiation of germanium, and to the annealing stages at ~35 and 65 K. But we will not touch on these issues.

It is well known that in *n*-type germanium, when irradiated with *γ*-quantum, electrons, protons, etc., the conductivity tends to its eigenvalue, and then undergoes *n-p*-conversion. Already in 1964, to describe the high-temperature annealing (273–450 K) of radiation defects in antimony-doped germanium, the concepts of vacancy migration and the formation of vacancy + antimony and divacancy + antimony complexes were used [133]. In subsequent studies, it was shown that the assumption of the formation of vacancy – donor complexes well explains the radiation effects upon irradiation of germanium [92, 134-143]. The assumption of the formation of neutral vacancy – donor complexes allowed the authors of [144] to propose an original methodology for determining the degree of compensation of a material. Attempts to determine the position of the levels of the vacancy + donor complex in the forbidden zone of germanium were unsuccessful. The existence of such a complex in germanium can be confirmed by the results of Ref. [145], which showed that defects in silicon annealing at 150–180°C (which corresponds to the annealing temperature of *E*-centers) are similar to defects in germanium annealing at 30– 50°C.

In [134, 151], the formation of more complex complexes is also possible — a donor impurity + two vacancies. The levels E_v+0.10 eV (phosphorus, arsenic), E_v + 0.12 eV (antimony), E_v + 0.16 eV (bismuth) are attributed to this defect (see, for example, [146]). The annealing temperature of these associations increases with an increase in the covalent radius of the donor atom. However, there is a discrepancy between the annealing temperature of the arsenic + divacancy complex determined from electrical measurements (150°C) and the annealing temperature of the arsenic + divacancy complex identified from EPR measurements (~210°C) [147].

There is no convincing evidence for the identification of simple defects (vacancies, divacancies, interstitial atoms of germanium), although one can speak of the mobility of vacancies at low temperatures [148]. The possibility of the formation of interstitial germanium atoms was reported in [142, 146]. They, in particular, propose a new mechanism for reducing the mobility of charge carriers in irradiated germanium, which takes into account the deformation around the interstitial atom [143]. One can also speak with confidence about the existence of a vacancy + oxygen complex in Germany [148-150]. The possibility of the formation of an interstitial impurity (copper and alloying) was reported in [151, 152]. The proof of the interaction of simple defects with atoms of a dopant, metals - gold, zinc, mercury, lithium was obtained in [153-157].

Germanium p-type conductivity, doped with "ordinary" impurity (gallium, indium), as a rule, is quite resistant to electrons and γ-quantum at room temperature. But if the same material is irradiated at 78 K, then defects are formed that can reversibly rebuild when heated and illuminated. Apparently, such defects were first noticed in [158, 159]. The authors of independently noted the influence of light on these defects [160, 161]. The defect does not contain a dopant [162, 163], but it contains vacancies [163]. It is assumed [163] that most likely the rebuilding center consists of intrinsic defects - vacancies and interstitial atoms with the dominant role of vacancies. The defect model proposed in [162] admits that this is a fairly separated positively charged pair of vacancy - interstitial atoms. A thorough study [164] concludes that low-temperature irradiation introduces defects whose concentration does not depend on the concentration of the dopant, as well as associations associated with the dopant, but their concentration is much lower. The authors of [165] also adhere to the model proposed in [162]. But there is a paper [166], the authors of which believe that the stage of annealing of 80–140 K is caused by the migration of vacancies to the impurity atom with the formation of vacancy + impurity atom associations, and the stage 220–270 K by the migration of impurity atoms in interstitial positions to impurity substitution atoms. Estimates are given for the migration energy at internodes for gallium (0.4 eV) and indium (0.7 eV).

In Germany, the main stages of annealing are the stages of approximately 150 ° C and, at high radiation doses, -250°C. But samples doped with gold restore their properties only after annealing at 500 ~ 530°C.

3.8. Radiation defects in binary compounds

While in monoatomic semiconductors there are two simplest defects — a vacancy and an interstitial atom, eight are in binary compounds — two types of vacancies and substitutions and four types of interstitial atoms (atom A surrounded by atoms A and B and atom B surrounded by atoms A and B) It should be remembered that the various spatial configurations of the interstitial atom (in the tetrahedral or hexagonal internodes, differently oriented dumbbell configurations, asymmetric dumbbell configurations of two, different atoms, etc.) are not identical, in a number of parameters and can be considered as different defects . It is necessary to take into account the interaction of simple defects with impurities. In addition, it should be noted that in binary compounds there is more: the concentration of uncontrolled impurities (possibly, with the exception of indium antimonide). Therefore, during irradiation due to a shift in the Fermi level, impurity levels can be manifested that are present in the sample before irradiation. Studies of the effect of radiation on the properties of binary compounds have probably already passed the stage of accumulation of data on the thresholds of defect formation, levels of defects and impurities observed after one or another processing of the crystal (radiation or thermal), characteristics of annealing, on the rates of introduction of defects during irradiation, on the correlation of concentration input levels, with impurity concentration. There are reviews (for example, [167,170], where attempts have been made to systematize the available experimental data on binary compounds.

We consider the radiation effects in binary compounds using the example of indium antimonide. The consequences of irradiation of InSb by electrons at 80 K were studied in [171-175]. Upon irradiation of *n*-InSb, the electron concentration decreases linearly with the irradiation dose, and the hole concentration in InSb of *p*-type conductivity increases [171]. The removal rate of charge carriers with an initial concentration $n_o = 1{,}5 \cdot 10^{16} cm^{-3}$ when irradiated with electrons with an energy of 1 MeV was 8.6 cm^{-1}.

Isochronous annealing of *n*- and *p*-type InSb irradiated with 1 MeV electrons at the temperature of liquid nitrogen is shown in Fig. 3.32. Carrier recovery occurs in at least five stages between 80 and 320 K: *I* - 90; *II*- 150; *III* – 175; *IV – 210; V* - 275. In *n*-type samples, the annealing temperature in stages *I – IV* decreased with decreasing initial electron concentration.

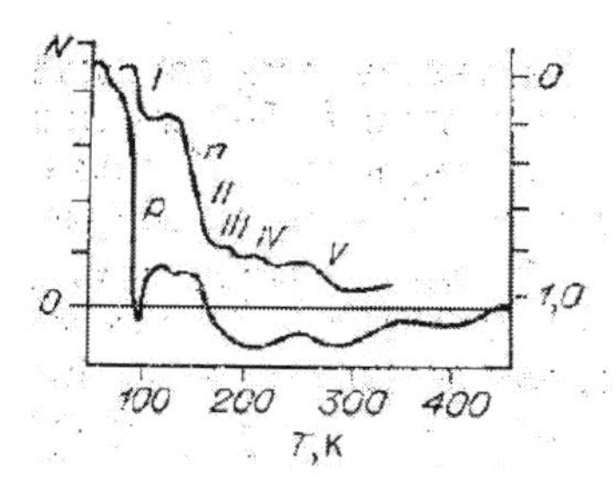

Figure 3.32. Stages of isochronous annealing of indium antimodine.

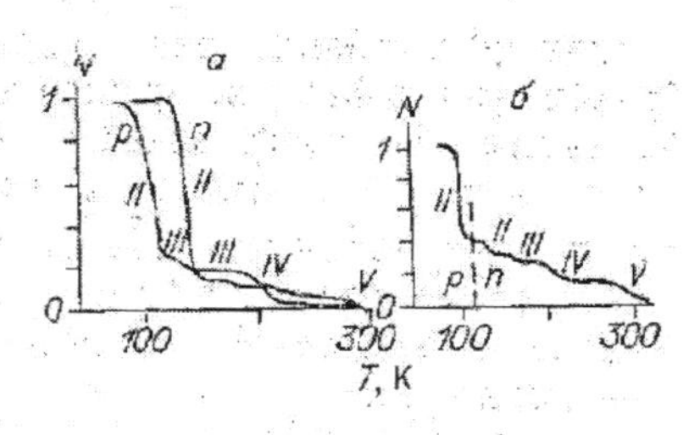

Figure 3.33. Five-minute isochronous annealing of InSb irradiated with γ-rays [178]. a) *p-n* no conversion; b) the *n*-type that has undergone conversion. Roman numerals are annealing stages.

The annealing temperature in stage *II* varied from 91 to 203 K with a change in the carrier concentration from 10^{16} (*p*-type) to 10^{18} (*n*-type), while the activation energy of annealing varies from 0.25 to 0.71 eV. When annealed *p*-InSb is annealed, several sections of reverse annealing appear [171].

In [175], *n*- and *p*-type InSb was irradiated with electrons with an energy of 4.5 MeV. The rate of removal of charge carriers depended on the position of the Fermi level, that is, on the initial concentration of charge carriers, and for *n*-type samples, the removal rate is noticeably higher than for *p*-type samples. When irradiated with *n*-InSb, the electron concentration drops to conductivity type conversions, after which the hole concentration increases. Upon irradiation of *p*-InSb, the hole concentration changes monotonously, but has a certain peculiarity: if the Hall coefficient $R_x>2000$ cm^3/C, then the hole concentration increases, otherwise ($R_x <2000$ cm^3/C) it decreases. Thus, upon irradiation, the Fermi level approaches a value that can be called the limiting Fermi level. Its position depends on the irradiation temperature: 0.03 eV for irradiation of about 80 K and 0.08 eV for irradiation of about 200 K. Three annealing stages were detected during annealing of irradiated *n*-InSb samples: *I* - 80-85; *II* - 120-150; *III* - 250-300 K. When annealing *p*-InSb, the same stages can be distinguished, although they are not so well defined (Fig. 3.33).

When irradiated at nitrogen and helium temperatures, the same carrier removal rates are observed [172], and it turned out that isochronous annealing after irradiation at helium temperature does not restore carriers to the temperature range corresponding to stage *I*. Irradiation at 165 K leads to the same complete removal of carriers, as if the sample was irradiated at 80 K and annealed about 165 K, and stages *III* – *V* of reduction take place. A similar result was obtained in [175] when irradiated at 200 K.

In order to determine the threshold energy for defect formation, we studied the orientation dependence of radiation damage in InSb [173]. One sample was irradiated in the <111> direction, and the other, <$\bar{1}\,\bar{1}\,\bar{1}$>. It was found that when the direction of irradiation is changed, the ratio of the number of defects annealed at different stages changes. This was expected from consideration of the crystal lattice of $A^{III}B^{V}$ compounds. The energy dependence of the rate of formation of defects annealed at stages *I* and *II* with two different directions of irradiation is obtained. It turned out that the rate of formation of defects annealed in stage *I* is higher during irradiation in the <111> direction, and the rate of formation of defects annealed in stage II is higher during irradiation in the <Ī Ī Ī> direction. Thus, it was concluded that the defects annealed in stage *I* are formed by the displacement of Sb atoms, and the defects annealed in stage *II* are formed by the displacement of In atoms. The threshold bias energies were determined: for indium - 6.4 eV, for antimony - 8.5-9.9 eV.

In studying stage *II* of annealing, it was concluded [171] that there are two types of defects responsible for this stage, and their energy levels were determined: E_1-E_v +0.04 eV, E_2-E_c -0.03 eV, both of which are acceptors. It is assumed that these are two different configurations of the interstitial complex - a vacancy due to the displacement of indium atoms during irradiation. All stages of restoration in [172] are interpreted as being due to the restoration of primary defects without noticeable formation of defect + impurity complexes.

In [175], the introduction of the level of E_v + 0.03 eV (apparently a donor) with a speed of 3.5 cm^{-1} was found in the case of irradiation at 80 K. Irradiation at 200 K introduces the levels of E_v + 0.048, E_v + 0.081, and a group of levels in the range from E_c - 0.03 to E_c - 0.10 eV with approximately the same injection rates of ~ 1.5 cm^{-1}. The author of this work suggests that the annealing features found between 80 and 200K cannot be explained only by the recombination of vacancies and internodes, since rearrangement of unannealed defects and a change in their charge state play an important role in annealing processes. Under γ irradiation of indium antimonide, as in the case of electron irradiation, a decrease in the concentration of electrons in the *n*-type material and an increase in the concentration of holes in the *p*-type material are observed [176]. After conversion of the conductivity type in n-InSb, the hole concentration grows and tends to a limiting value that is little dependent on the initial electron concentration. In *p*-type InSb, the concentration of holes upon irradiation also tends to a limiting value. In the studied samples, the limiting hole concentrations for *n*- and *p*-type InSb are close to each other ($n_0 = 2{,}9 \cdot 10^{13}$ - $1{,}6 \cdot 10^{14}$, $p_0 = 5{,}3 \cdot 10^{13}$-$3{,}3 \cdot 10^{14}$) [176]. Initial radiation defect formation rates $\Delta n/\Delta\Phi$ and $\Delta p/\Delta\Phi$, defined in [176], are independent of the initial concentration of charge carriers and are close to *n*- and *p*- InSb ($\Delta n/\Delta\Phi \sim 5{,}8.10^{-3}$; $\Delta p/\Delta\Phi \sim (6{,}3 - 8{,}8)10^{-3}$).

During isochronous annealing of γ-irradiated *p*- and *n*- InSb (see Fig.

3.33a), four stages of reduction are found [178], corresponding to stages *II* – *V* observed upon electron irradiation. According to the authors, stage *I* is absent because the defects responsible for it are annealed during irradiation. Stage *II* cleavage was observed for *n*-InSb converted to *p*-type by irradiation, and a sharp increase in the activation energy of the process (from 0.24 to 0.40 eV). Analyzing the totality of data on the irradiation and annealing of InSb by electrons of various energies (from 0.4 to 4.5 MeV) and γ-rays, the authors of [178] note that there is no dependence of the position of the annealing stages on the radiation energy. On the other hand, the ratio of the concentration of defects annealed at different stages depends on the radiation energy: the higher the energy, the greater the contribution of stages *III* - *V*.

In [176, 177], the energy levels of defects arising in indium antimonide during γ-irradiation and annealing were studied. Three levels, apparently belonging to radiation defects, are distinguished: 1) $E_v + 0.038$ eV is formed immediately after *n*- *p*- conversion, the rate of formation is close to the initial rate of electron decrease; 2) $E_c - 0.083$ eV - donor; 3) $E_v - 0.048$ eV - acceptor, manifests itself upon irradiation and anneals at stage *I*.

In [179], it was found that, upon irradiation of *n*-InSb, immediately after *n*-*p*- conversion, the concentration of M_2 acceptors with a level of $E_v + 0.040$ eV increases. In the *p*-type material doped with Ge and Mn, immediately after irradiation, the concentrations of M_3 and M_4 acceptor states $E_v + 0.034$ eV begin to increase. The concentrations of M_2, M_3, and M_4 increase linearly with the dose and reach limit values close to the concentrations of impurities — donors in the *n*-type material, Ge and Mn in the *p*-type material. Based on the results obtained, the authors of [179] believe that the defects M_2, M_3, and M_4 include impurity atoms, and this is the basis for assuming the nature of these defects.

When InSb was irradiated with X-rays with a quantum energy below the threshold ($h\nu$ = 55 [178, 180], 100 [181] and 250 keV [182, 183]) at a temperature of 78 K, radiation defects also appeared. In *p*-type samples, the hole concentration, conductivity, and mobility in the same direction significantly changed as in the case of high-energy γ-irradiation. The electron concentration in *n*-InSb did not change, however, after preliminary irradiation with γ-rays and annealing to room temperature, the electron concentration decreased markedly when irradiated with X-rays. In *p*-InSb upon repeated irradiation, the initial rate of defect formation is greater than during the first irradiation [180]. The initial rate of hole introduction into *p*-InSb, determined in [178], turned out to be $(4\pm2)\cdot10^{-5}$ cm^{-1}. It was established in [182] that the limiting change in the hole concentration $\Delta p\sim1{,}2\cdot10^{14}$ is the same for all samples ($p_0=10^{12}$-10^{15}). The concentrations of N_a acceptors and N_d donors at the same rate decreased upon irradiation of InSb doped with Ge and Mn.

Annealing of the changes takes place within one stage, regardless of the initial carrier concentration at 95-100 K [178, 180, 182, 183]. Annealing can also be performed using a CO_2 laser while maintaining the sample temperature not higher than 80 K. Authors [178] It is believed that the formation of defects during x-ray irradiation is associated with the action of the Varley mechanism. In [183], the possibility of creating defects using the ionization mechanism was considered in detail.

We studied the behavior of InSb upon irradiation with light from the intrinsic absorption region at 78 K (DKS-1000 lamp) without a filter and with a Ge filter [184]. The initial rate of change in the hole concentration in p-InSb was ~ 2.10^{-9} $cm^{-1}\cdot eV^{-1}$. In *n*-InSb, as with x-ray irradiation, no changes were observed. Defects are annealed at 100 K. It is shown that when irradiated with light, defects are created in the volume of a thin surface layer, the thickness of which is ≤ 30 μm.

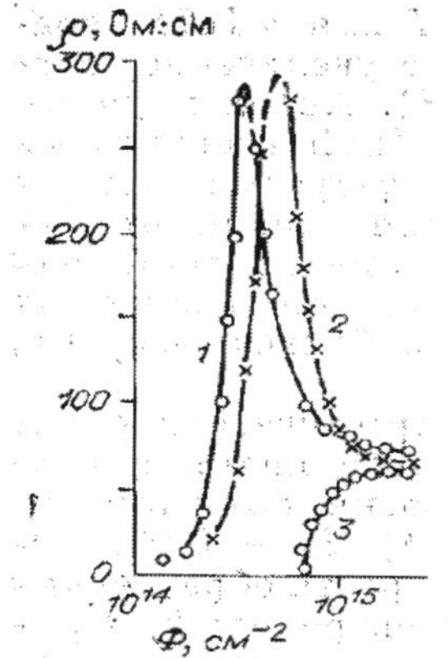

Figure 3.34. Dependence of the resistivity of *n*- and *p*-type indium antimonide samples on the integrated dose of fast neutrons [186].
Specific resistance at T = 77 K, Ohm.cm: *1* - 6.4; *2* - 3.4; *3* - 0.07 (1.2 - *p*-type samples, 3-*n*-type)

When irradiating InSb crystals at 78 [186] and 300 K [185, 187] with fast neutrons, in contrast to electron irradiation, *p*-type samples undergo a conductivity type conversion, and in *n*-type samples the conductivity decreases (Fig. 3.34). Isochronous annealing of InSb samples irradiated at 78 K reveals two annealing stages: 77–150 and 150–260 K. After annealing to room temperature, the carrier concentration is restored; however, the resistivity and mobility are not completely annealed; therefore, some radiation defects remain stable. playing the role of scattering centers.

In [187], the effect of neutron irradiation at 300 K on the mobility of holes in *p*-InSb crystals was studied. The temperature dependence of mobility before and after irradiation with a dose of $1 \cdot 10^{17}$ cm^{-2} is shown in Fig. 3.35. When irradiated with a dose of $1 \cdot 10^{16}$ cm^{-2}, this dependence was not observed (although the mobility decreased). The authors of [187] suggest that the defects responsible for the decrease in mobility are centers of the type of disordered *n*-type regions with a double electric layer. In [188], the effect of neutron irradiation on electron mobility in plastically deformed *n*-InSb crystals was studied. In unirradiated deformed samples, mobility sharply decreased at 77 K. After irradiation with fast neutrons, a decrease in mobility was observed at 77 K in samples with a low degree of deformation, with a greater degree of deformation (ε>0.5%), the mobility increased

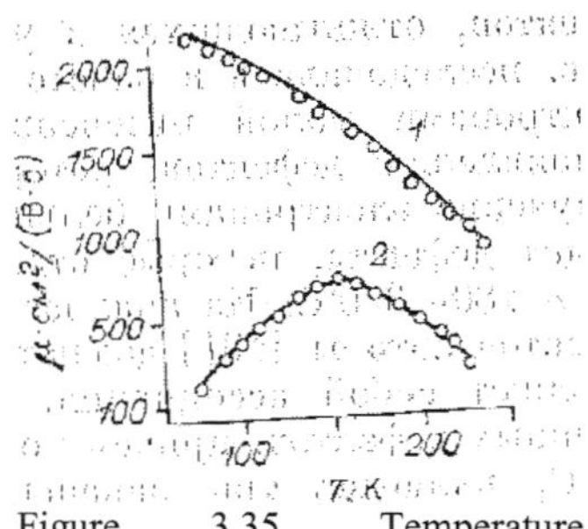

Figure 3.35. Temperature dependence of hole mobility in InSb crystals [187].
1 - before irradiation; 2 - after irradiation with an integrated fast neutron flux of $1 \cdot 10^{17}$ cm^{2} at 300 K.

The dislocation density after irradiation did not change. It was assumed in [188] that scattering in irradiated crystals containing a large number of dislocations $\sim 4 \cdot 10^{6}$, $2 \cdot 10^{7}$ is caused by the association of radiation defects with dislocations. This may be a defect as a result of the interaction of the positive space charge of the impurity atmosphere of the dislocations with the negative external electric field of the disordered regions.

The authors of [189] found that upon bombardment of *p*-InSb with protons, an *n*-type layer forms near the surface. Irradiation was carried out by protons with an energy of 100 keV at T_{room}, the integral dose of 10^{14} cm^{-2}. A 1 μm thick *n*-layer is formed with an average carrier concentration in the layer of $\sim 10^{17}$ cm^{-3} (the initial concentration of holes in the *p*-type material is approximately $5 \cdot 10^{15}$ cm^{-3} at 77 K). Based on the *p-n*-junction created by proton bombardment, the authors of [189] fabricated a photodiode. At 77 K, the zero bias resistance of the diodes was several hundred thousand ohms. The maximum detectability of these diodes at 4.9 μm was more than $3 \cdot 10^{10}$ cm Hz$^{1/2}$/W with the largest observed value of 10^{11} cm·Hz$^{1/2}$/W (with a background environment of 300 K). The efficiency of diodes is about 35%.

The properties of the bombarded layer did not change during 460-hour annealing at 80°C. After irradiation at room temperature, two annealing stages are observed [190] in the temperature ranges 60–180 and 180–330°C. The role of intensity upon irradiation with 7 MeV electrons on the accumulation of defects annealing in the indicated temperature range was studied in [191]. It was shown that irradiation with low-intensity electrons ($3{,}2 \cdot 10^{12}$ cm^{-3}) leads to the appearance of defects that anneal at 60 - 180°C. Irradiation with high-intensity electrons ($16 \cdot 10^{12}$ cm^{-3})] introduces defects that are annealed in two stages: at 60 - 180 and 180 - 330°C. The authors of [191] conclude from the analysis of the kinetics of accumulation and annealing of defects: the introduced defects are associations consisting of the simplest primary defects, and the complex annealed at 60–180°C contains three vacancies (or interstitial atoms), while the more stable complex contains no less than five vacancies.

References:

1. McKinley A., Feshbach H. The Coulomb Scattering of Relativistic Electrons by Nuclei.- Phys. Rev., 1948, v. 74, N 12, p. 1759 - 1963.
2. Mecci G. *Teoriia atomnykh stolknovenii* [Atomic collision theory]. Moscow. Mir. 1969. 756 p. (in Russian)
3. Kinchin G.N., Piz R.S. *Smeshchenie atomov tverdykh tel pod deistviem izlucheniia* [Displacement of atoms of solids under the influence of radiation]. Usp. fiz. Nauk. 1956. Vol. 60, iss. 4. p. 590—615. (in Russian)
4. Cahn J. H. Irradiation Damage in Germanium and Silicon Due to Electrons and Gamma Rays. J. Appl. Phys., 1959, Vol. 30, p. 1310-1316.
5. Ivanov H.A., Kosmach V.F., Kulakov V.M. and other. *Vklad produktov iadernykh reaktsii v obrazovanie prostykh defektov pri obluchenii elektronami i γ-kvantami* [Contribution of nuclear reaction products to the formation of simple defects when irradiated with electrons and γ-quanta]. Fiz. i tekhn. poluprov. 1975. Vol 9, Iss. 6, p. 1049-1052. (in Russian)
6. Kelli B. *Radiatsionnoe povrezhdenie tverdykh tel* [Radiation damage to solids]. Moscow. Atomizdat. 1970. 236 p. (in Russian)

7. Vavilov V.S. *Deistvie izluchenii na poluprovodniki* [The effect of radiation on semiconductors]. Moscow. Fizmat-giz. 1963. 264 p (in Russian)
8. Bulgakov Iu.V., Kumakhov M.A. *Prostranstvennoe raspredelenie radiatsionnykh defektov v materialakh, obluchennykh puchkami mono- energeticheskikh chastits* [Spatial distribution of radiation defects in materials irradiated with beams of monoenergetic particles]. Fiz. i tekhn. poluprov. 1968. Vol. 2, Iss. 11, p. 1603-1607. (in Russian)
9. Konopleva R.F. Ostroumov V. I. *Vzaimodeistvie zariazhennykh chastits vysokikh energii s germaniem i kremniem* [The interaction of high-energy charged particles with germanium and silicon]. Moscow. Atomizdat. 1975. 128 p. (in Russian)
10. Kosmach V.F., Kulakov V.M., Ostroumov V.I., Petukhov A.M. *Raschet chisla defektov, obrazovannykh v kremnii bystrymi neitronami* [Calculation of the number of defects formed in silicon by fast neutrons]. Fiz. i tekhn. poluprov. 1972. Vol. 6, Iss. 2, p. 420-422. (in Russian)
11. Radiation Effects in Semiconductors. N. Y. Plenum Press. 1968.
12. Radiation Effects in Semiconductors. N. Y. Gordon and Breach. 1971.
13. Proceeding of the International Conference on Radiation Damage and Defects in Semiconductors. London. Institute of Physics. 1973.
14. Lattice Defects in Semiconductors. Conf. Ser. N 23. Inst. of Phys. London - Bristol. 1975.
15. Radiation Effects in Semiconductors. Conf. Ser. N 31. Inst. of Phvs. Bristol – London. 1976.
16. Vavilov V.S., Ukhin N.A. *Radiatsionnye effekty v poluprovodnikakh i poluprovodnikovykh priborakh* [Radiation effects in semiconductors and semiconductor devices]. Moscow. Atomizdat. 1969. 311 p. (in Russian)
17. Corbett J. W. Electron Radiation Damage in Semiconductors and Metals. N. Y. Acad. Press. 1966.
18. Konozenko I.D., Semeniuk A.K., Khivrich V.I. *Radiatsionnye defekty v kremnii* [Silicon radiation defects]. Kiev. Naukova dumka. 1974. 220 p. (in Russian)
19. *Fizicheskie protsessy v obluchennykh poluprovodnikakh* [Physical processes in irradiated semiconductors]. Novosibirsk. Nauka. 1977.

256 p. (in Russian)
20. Konopleva P.F., Litvinov V.L., Ukhin N.A. *Osobennosti radiatsionnnogo povrezhdeniia poluprovodnikov chastitsami vysokikh energii* [Features of radiation damage to semiconductors by high-energy particles]. Moscow. Atomizdat. 1971. 176 p. (in Russian)
21. Watkins G.D., Corbett J.W. Defects in Irradiated Silicon: Electron Paramagnetic Resonance of Divacancy. Phys. Rev. 1965. Vol. 138, № 2A, p. 548 – 555.
22. Daly D.F., Noffke H.E. An EPK Study of Fast-Neutron Radiation Damage in Silicon. - In: Radiation Effects in Semiconductors. London – N. Y. - Paris, Gordon and Breach. 1971, p. 179 - 188.
23. Barnes C.E. Absorption Measurements in Neutron Irradiated Silicon. In: Radiation Effects in Semiconductors. London - N.Y. - Paris, Gordon and Breach. 1971, p. 203 - 210.
24. Cheng L.J., Corelli J.C., Corbett J.W., Watkins G.D. 1,8-, 3,3-and 3,9 μ Bands in Irradiated Silicon: Correlation with the divacancy. Phys. Rev. 1966, Vol. 152, Iss. 2, p. 761-774
25. Lee Y.H., Brosions P.R., Corbett J.W. New EPR Spectra m Neutron Irradiated Silicon (II). Rad. Effects, 1974. Vol. 22, p. 139 - 172.
26. Jung W., Newell G.S. Spin-1 Centers in Neutron Irradiated Silicon. Phys. Rev. 1963. Vol. 132, Iss. 2, p. 648 - 662.
27. Brower K.L., Vook F.L., Borders J.A. Electron Paramagnetic Resonance of Defects in Ion Implanted Silicon. Appl. Phys. Lett., 9196, Vol. 15, Iss 7, p. 208 - 210
28. Brower K.L., Beezhold W. Electron Paramagnetic Resonance of the Lattice Damage in Oxygen-Implanted Silicon. J. Appl. Phys., 1972, Vol. 43, Iss 8, p. 3499 - 3506.
29. Lee Y. H., Corbett J. W. EPR studies in Neutron Irradiated Silicon: A Negative Charge State of a Non-Planar Five-Vacancy Cluster (r_s). Phys. Rev. 1973. Vol. B 8, p. 2810 - 2826.
30. Lee Y.H., Corbett J.W. EPR Study of Defects in Neutron Irradiated Silicon: Quenched in Alignment under 110 - uniaxial Stress. Phys. Rev., 1974. Vol. 9, Iss. 10, p. 4351 – 4361.
31. Nisenoff M., Fan H.I. Electron Spin Resonance in Neutron Irradiated Silicon. Phys. Rev. Vol. 128, Iss 4, p. 1605 – 1613.
32. Corbett J.W., Wotkins G.D., Silicon Divacancy and Its Direct

Production by Electron Irradiation. Phys. Rev. Lett. 1961, Vol.7, Iss 8, p.314-316

33. Vavilov B.C., Isaev H. U., Mukashev B. H., Spitsin A. V. *Bliianie razmera atomov donornoi primesi na nakoplenie i otzhig radiatsionnykh defektov v kremnii n-tipa* [Effect of the size of atoms of a donor impurity on the accumulation and annealing of radiation defects in *n*-type silicon]. In book: *Radiatsionnye defekty v poluprovodnikakh* [Radiation defects in semiconductors]. Minsk, Publishing House of the Belarusian State University named after Lenin. 1972, p.15 – 17. (in Russian)
34. Hasiguti R. R., Ishino S., Defect Mobility and Annealing in Irradiated Germanium und Silicon. In: Radiation Damage in Semiconductors. Paris, Dunod. 1965, p. 15 -17
35. Wotkins G.D. Defects in Irradiated Silicon: Electron Paramagnetic Resonance and Electron-Nuclear Doule Resonance of the Alummmm – Vacancy Pair. Phys. Rev., 1967, Vol. 155, p. 802 – 815.
36. Wotkins G.D. EPR of a Trapped Vacancy in Boron – Doped Silicon. Phys. Rev., B, 1976. Vol. 13, Iss 6, p. 2511 – 2518.
37. Wotkins G.D., Defects in Irradiated Silicon: EPR of the Tin – Vacancy Pair. Phys. Rev., B, 1975. Vol. 12, Iss. 10, p. 4383 – 4390.
38. Wotkins G.D. Microscopic View of Radiation Damage Semiconductors Using EPR as a Probe. IEEE Trans. Nucl. Sci., 1969, v. NS-16 p. 13 – 18.
39. Bemski G. Paramagnetic Resonance in Electron Irradiated Silicon J. Appl. Phys., 1959, v. 30, p. 1195 - 1198; Watkins G. D. e. a., ibid., p. 1198 – 1199.
40. Watkins G.D. Review of EPR Stafe in Irradiated Silicon: Radiation Damage in Semiconductors, Paris. Dunod.1965. p. 97 – 113.
41. Brower K.L. Electron Paramagnetic Resonance of Neutral (s=1) One Vacany- Oxygen Center in Irradiated Silicon. Phys. Rev., 1971. Vol. 4, Iss 6, p. 1698 – 1782.
42. Brower K.L. Structure of Multiple - Vacancy (Oxygen) Centers in Irradiation Silicon. Rad. Eff., 1971. Vol. 8, Iss 3, 4, p.2I3 – 219.
43. Lappo M.T. *IK – pogloshchenie na ostatochnykh primesiakh, termo- i radiatsionnykh defektakh v kremnii* [IR - absorption on residual impurities, thermal and radiation defects in silicon]. Abstract of

candidate dissertation. Minsk. 1973, 11 p. (in Russian)

44. James H.M., Lark-Horovitz H. Localized Electronic States in Bombarded Semiconductors. J. Phys. Chem., 1951, Vol. 198, p. 107 – 115.
45. Blount E.J. Energy Levels in Irradiated Germanium. J. Appl. Phys. 1959, Vol. 30, Iss. 8, p.1218 - 1221
46. Gwozdz P. S., Koehler J.S. Changes in ac Conductivity of Silicon with Electron Irradiation et 0,5 K. Phys. Rev., 1972, Vol. 6, p. 4571 – 4574.
47. Berry B.S., Frank W., Tan S.J. In: Ion Implantation in Semiconductors and other Materials. N.Y., Plenum Press, 1973, p. 19 - 25.
48. Frank W, Berry B.S. Lattice Location and Atomic Mobility of Implanted B in Si. Rad. Eff., 1974, Vol. 21, Iss 2, p. 105 – 111.
49. Frank W. The Nature of Interstitials in Silicon and Germanium. in: Lattice Defects Semiconductors, Conf. Ser. N 23, the Inst. of Phys., London – Pristol, 1975, p. 23 – 43.
50. Lee J.H., Corbett J.W. EPR Evidence of the Self - Interstitials in Neutron Irradiated Silicon. Sol. St. Com.1974, Vol.15, Iss 11/12 p. 1781- 1784.
51. Lee Y.H., Gerasimenko N.N., Corbett J.W. An EPR Study of Neutron - Irradiated Silicon: A Positive Charge of the <100>. Split Di-Interstitial. Phys. Rev., 1976, Vol. B 14, p. 4506-4517.
52. Brower K. L. EPR K of a <001> Si Interstitial Complex in Irradiated Silicon. Phys. Rev., 1976, Vol. B 14, p. 872 – 882.
53. Lee J. H., Kim Y. M., Corbett J. W. New EPR Spectra in Neutron Irradiated Silicon. Rad. Eff., 1972, Vol.15, p. 77 – 84.
54. Botvin V.A., Gorelkinsky Yu.V., Sigle. V.O., Chubisov M.A., *Paramagnitnye tsentry v kremnii, obluchennom tiazhelymi zariazhennymi chastitsami* [Paramagnetic centers in silicon irradiated by heavy charged particles]. Semiconductor Physics and Technology. 1972. Vol.6, Iss.9, p. 1683 – 1686. (in Russian)
55. Brower K.L. Electron Paramagnetic Resonance of the Aluminium Interstitial in Silicon. Phys. Rev., 1970, Vol. B 1, p.1908 – 1917.
56. Bean A.R, Newman R.C., Smith R. S. Electron Irradiation Damage in Silicon Containing Carbon and Oxygen. J. Phys Chem. Sol., 1970, Vol. 31, Iss 4, p. 739 – 751.

57. Broziel A.R., Newman R.C., Totterbell D.H. Interstitial Defects Involving Carbon in Irradiated Silicon. J. Phys., C: Solid State Phys., 1975, Vol.8, p. 243 - 248.
58. Lee J.H., Corbett J.W., Brover K.L., EPR of Garhon - Oxygen - Divacancy Complex in Irradiated Silicon; Phys. Stat. Sol. (a), 1977, Vol.41, p. 637-647.
59. Gossik B.R. Disordered Regions in Semiconductors Bombarded by Fast Neutrons. J. Appl. Phys., 1959, Vol.30, Iss 8, p 1214 - 1219.
60. Ukhin H.A. *Model' razuporiadochennykh oblastei v kremnii, sozdavaemykh bystrymi neitronami* [Model of disordered regions in silicon created by fast neutrons]. Semiconductor Physics and Technology. 1972, Vol.6, Iss 5, p. 931 - 934. (in Russian)
61. Lenchenko V.M., Akilov Iu.Z. *Issledovanie s pomoshch'iu EVM struktury kaskadov smeshchenii v Ge, Si i RbS* [Computer study of the structure of cascades of displacements in Ge, Si, and RbS]. Semiconductor Physics and Technology. 1971. Vol.5, Iss. 3, p. 397 – 402. (in Russian)
62. Akilov Iu.Z., Lenchenko V.M. *Kaskady smeshchenii atomov v Ge i Si (mashinnoe modelirovanie)* [Cascades of atomic displacements in Ge and Si (machine simulation)]. Semiconductor Physics and Technology. 1974. Vol.8, Iss. 1, p. 30 – 38. (in Russian)
63. Martynenko Yu.V. Annealing and Clustering of Defects in Cascades. Rad. Eff. 1976. Vol. 29, Iss 3, p. 129 – 135.
64. Baranov A.I. *Nakoplenie defektov i protsessy amorfizatsii pri bombardirovke poluprovodnikov ionami* [Accumulation of defects and amorphization processes during the bombardment of semiconductors by ions.]. In book: Radiatsionnye effekty v poluprovodnikakh [Radiation effects in semiconductors]. Novosibirsk. Science. 1979. (in Russian)
65. Baranov A.I., Smirnov L.S. *O vzaimodeistvii razuporiadochennykh oblastei i okruzheniia v poluprovodnikakh* [On the interaction of disordered regions and the environment in semiconductors]. Semiconductor Physics and Technology. 1973, Vol. 7, Iss. 11, p 2227 – 2229. (in Russian)
66. Gerasimenko N.N., Kibalina N.P., Stas' V.F. *Pripoverkhnostnyi sloi v obluchennom kremnii* [Surface layer in irradiated silicon]. In book:

Radiatsionnye effekty v poluprovodnikakh [Radiation effects in semiconductors]. Novosibirsk. Science. 1979. (in Russian)
67.Smirnov L.S., Tishkovskii E.G. *Otsenka vremeni zhizni svobodnykh vakansii v kristallakh kremniia* [Estimation of the lifetime of free vacancies in silicon crystals]. Semiconductor Physics and Technology. 1978. Vol. 12, Iss. 3, c. 543 – 548. (in Russian)
68.Khainovskaia V.V., Smirnov L.S. *Vzaimodeistvie radiatsionnykh defektov s dislokatsiiami v germanii* [Interaction of radiation defects with dislocations in germanium]. Solid state physics, 1967, Vol. 9, Iss. 7, p. 2043 – 2046. (in Russian)
69.Baranov A.I., Gadniak G.V., Ruzankin S.F. *Chislennoe modelirovanie defektov* [Numerical simulation of defects]. Sb. VIMI «Riport», 1975, № 16.
70.Morozov N.P., Tetel'baum D.I., Pavlov P.V., Zorin E.I. *Diffuzionno-koaguliatsionnaia model' nakopleniia radiatsionnykh defektov pri ionnoi bombardirovke* [Diffusion-coagulation model of the accumulation of radiation defects during ion bombardment]. Semiconductor Physics and Technology. 1975. Vol. 9, Iss. 12, p. 2292-2295. (in Russian)
71.Morozov N.P. Tetel'baum D.I., Pavlov P.V., Zorin E.I. The Calculation of Secondary Defect Fomation at Ion Implantation of Silicon. Phys. Stat. Sol. (a), 1976, Vol. 37, Iss 1, p. 57 – 64.
72.Madden P. K., Davidson S. M. The Nature of Rod-Like Defects Observed in B^+ Irradiated Si. Rad. Eff., 1972, Vol.14, Iss 3-4, p. 271-277.
73.Seshan K., Washburn J. The Nature and Habit Planes of Defekts in P lon Implantated Si. Phys. Stat. Sol. (a), 1974, Vol. 26, Iss 1, p. 345 - 352.
74.Pcheliakov O.P., Aseev A.L., Smirnov L.S., Stenin S.I. *Radiatsionnye vysokotemperaturnye effekty v germanii pri obluchenii v vysokovol'tnom elektronnom mikroskope* [Radiation high-temperature effects in germanium when irradiated in a high-voltage electron microscope]. Semiconductor Physics and Technology. 1976. Vol.10, Iss.8, p. 1472 – 1479. (in Russian)
75.Rodes R.G. *Nesovershenstva i aktivnye tsentry v poluprovodnikakh* [Imperfections and active centers in semiconductors]. Moscow.

Metallurgy. 1968. 371 p. (in Russian)

76. Boltaks B.I. *Diffuziia i tochechnye defekty v poluprovodnikakh* [Diffusion and point defects in semiconductors]. St. Petersburg. The science. 1972. 384 p. (in Russian)

77. Konorova L.F. *Termicheskie defekty v germanii* [Thermal defects in germanium]. Solid state physics. 1968. Vol.10, Iss. 9, p.2831-2834. (in Russian)

78. Boltaks B.I., Konorova L.F. *Vliianie termicheskikh defektov na vremia zhizni nositelei toka v germanii* [Influence of thermal defects for the lifetime of the current carriers in germanium]. Semiconductor Physics and Technology. 1970, Vol.4, Iss.4, p. 754–759. (in Russian)

79. Abdurakhmanova P.N., Vitovskii N.A., Maksimov M., Mashovets T.V. *Issledovanie termodefektov v germanii vysokoi chistoty*. [The study of thermal defects in high purity germanium]. Semiconductor Physics and Technology. 1970, Vol.4, Iss.12, p. 2298 – 2305. (in Russian)

80. Mashovets T.V., Ryvkin P.M., Khansevarov R. Iu. *O vliianii termo defektov na radiatsionnuiu stoikost' germaniia* [On the effect of thermal defects on radiation resistance of germanium]. Solid state physics. 1967. Vol.9, Iss. 2, p. 535 - 538. (in Russian)

81. Baranskii P.I., Klochkov V.P., Potykevich N.V. *Poluprovodnikovaia elektronika* [Semiconductor electronics]. Kiev. Naukova dumka. 1975. 704 p. (in Russian)

82. Kurilo P.M., Seitov E., Khitren' M.I. *Vliianie termicheskoi obrabotki na elektricheskie svoistva p-kremniia, soderzhashchego vysokuiu kontsentratsiiu kisloroda* [The effect of heat treatment on the electrical properties of n-silicon containing a high oxygen concentration]. Semiconductor Physics and Technology. 1970. Vol. 4, Iss. 12, p. 2267 – 2270. (in Russian)

83. Bibik V.F., Dadykin AA., Titov V.A. *Vliianie vysokotemperaturnogo progreva v vakuume na pripoverkhnostnyi sloi kremnievykh kristallov* [The effect of high-temperature heating in vacuum on the surface layer of silicon crystals]. Ukr. Physical Journal. 1975. Iss.10, p. 1684 – 1688. (in Russian)

84. Gartseva L.E., Levchuk L.V., Mordkovich V.N., Starchik M.N. *Povedenie kremniia, soderzhashchego kislorod, pri odnovremennykh*

termoobrabotke i obluchenii [The behavior of silicon containing oxygen, while heat treatment and irradiation.]. In book: *Radiatsionnaia fizika nemetallicheskikh kristallov* [Radiation physics of non-metallic crystals]. Vol.3. Part 1, Kiev. Naukova dumka. 1971. p. 284 – 288. (in Russian)

85.Bortnik M.V., Novokreshchenskaia T.I., Iukhnevich A.V. *Radiatsionnye narusheniia v termoobrabotannykh kristallakh kremniia, obluchennykh γ-kvantami* So^{60} [Radiation disturbances in heat-treated silicon crystals, irradiated with γ-rays of Co^{60}]. In book: *Radiatsionnaia fizika nemetallicheskikh kristallov* [Radiation physics of non-metallic crystals] Vol. 3. Part 4. Kiev. Naukova dumka. 1971. p. 264-269. (in Russian)

86.Panov V.I., Smirnov L.P., Tishkovskii E.G. *O vzaimodeistvii termo- i radiatsionnykh defektov v kremnii* [On the interaction of thermal and radiation defects in silicon]. Semiconductor Physics and Technology. 1975. Vol.9, Iss.8, p. 1580 – 1583. (in Russian)

87.Khainovskaia V.V., Smirnov L.P. *O vzaimodeistvii radiatsionnykh defektov s dislokatsiiami v germanii* [On the interaction of radiation defects with dislocations in germanium]. Semiconductor Physics and Technology. 1966. Vol.8, Iss.12, p. 3403 – 3404. (in Russian)

88.Khainovskaia V.V., Smirnov L.P. *O vzaimodeistvii radiatsionnykh defektov s dislokatsiiami v germanii* [On the interaction of radiation defects with dislocations in germanium]. Semiconductor Physics and Technology. 1967, Vol.9, Iss.7, p. 2043 – 2046. (in Russian)

89.Khainovskaia V.V., Smirnov L.P. *O vzaimodeistvii radiatsionnykh defektov s dislokatsiiami v germanii* [On the interaction of radiation defects with dislocations in germanium]. Semiconductor Physics and Technology. 1968, Vol.10, Iss.6, p. 1549 – 1550. (in Russian)

90.Smirnov L. P., Stas' V.F., Khainovskaia V.V. *Obluchenie i otzhig germaniia s bol'shoi plotnost'iu dislokatsii* [Irradiation and annealing of germanium with a high dislocation density]. Semiconductor Physics and Technology. 1969. Vol.3, Iss.12, p. 1760 – 1765. (in Russian)

91.Fridel' Zh. *Dislokatsii* [Dislocations]. Moscow. World. 1967. p. 643. (in Russian)

92.Smirnov L.P., Stas' V.F., Khainovskaia V.V. *Vliianie dislokatsii na kinetiku nakopleniia radiatsionnykh defektov v germanii* [The effect of dislocations on the kinetics of the accumulation of radiation defects in germanium.]. Semiconductor Physics and Technology. 1971. Vol.5, Iss.1, p. 85 – 90. (in Russian)

93. Romanov P.I., Smirnov L.P. *O vzaimodeistvii tochechnykh defektov s granitsei razdela SiO_2 – Si* [On the interaction of point defects with a SiO_2 - Si interface]. Semiconductor Physics and Technology. 1976. Vol.10, Iss.5, p. 876 – 881. (in Russian)
94. Romanov P.I., Smirnov L.P. *Obrazovanie razuporiadochennykh sloev pri bombardirovke kristallov ionami* [The formation of disordered layers during ion bombardment of crystals]. Semiconductor Physics and Technology. 1972. Vol.6, Iss.8, p. 1631 – 1634. (in Russian)
95. Gerasimenko N.N., Dvurechenskii A.V., Kachurin G.A. and other. *Radiatsionnyi otzhig defektov, obrazuiushchikhsia pri bombardirovke kristallov ionami* [Radiation annealing of defects formed during the bombardment of crystals by ions]. Semiconductor Physics and Technology. 1972, Vol.6, Iss.9, p. 1834 – 1835. (in Russian)
96. Mordkovich V.N., Solov'ev P.P., Temper E. M., Kharchenko V.A. *O radiatsionnom otzhige defektov v obluchennom neitronami kremnii* [On radiation annealing of defects in neutron-irradiated silicon]. Semiconductor Physics and Technology. 1974. Vol.8, Iss.5, p. 1024 – 1025. (in Russian)
97. Vikhrev B.I., Gerasimenko N.N., Dvurechenskii A.V., Smirnov L.P. *Vzaimodeistvie v kremnii atomov vodoroda s defektami, vvedennymi ionnoi bombardirovkoi* [Interaction of hydrogen atoms in silicon with defects introduced by ion bombardment]. Semiconductor Physics and Technology. 1974, Vol.8, Iss. 7. p. 1345 - 1348. (in Russian)
98. Sander H.H., Gregory B.L. Transient Annealing in Semiconductors Devices Following pulsed Neutron Irradiation. IEEE Trans. Nucl. Sci., 1966, v. NS-13, p, 53-62.
99. Binder D., Butcher D.T., Crepps J.R., Hammer E.L. Rapid Annealing in silicon Transistors. IEEE Trans. Nucl, Sci., 1968. Vol. NS-15, p.84-87.
100. Gregory B.L., Sander H.H. Transient. Annealing of Defects in Irradiated Silicon Devices. Proc. IEEE, 1970, Vol. 58, Iss 9, p. 1328-1342.

101.Srour J. R., Curtis O. L. Short-Term Annealing in Silicon Devices Folloving 14 Mev Neutron Irradiation. IEEE Trans. Nucl, Sci., 1972. Vol. 20 Iss 6, p. 196 - 203.

102.Tanaka T., Inuishi J. Hall Effect Messurement of Radiation Damaged and Annealing in Si. J. Phys. Soc. Japan, 1964, Vol. 19, Iss 2, p. 167 - 174.

103.Vologdin E. N., Zhukova G. A., Mordkovich V. H. *Obluchenie kremniia, pokrytogo okisnoi plenkoi, zariazhennymi chastitsami nizkikh energii* [Irradiation of silicon coated with an oxide film, charged particles of low energy]. Semiconductor Physics and Technology. 1972. Vol. 6, Iss. 7, p. 1306 – 1309. (in Russian)

104.Smirnov L.P., Stas' V.F., Khainovskii V.V. *Rol' dislokatsii v protsesse otzhiga obluchennogo germaniia* [The role of dislocations during the annealing of irradiated germanium]. Semiconductor Physics and Technology. 1971, Vol. 5, вып, 6, p. 1179 - 1184. (in Russian)

105.Meier Dzh., Erikson L., Devis Dzh. *Ionnoe legirovanie poluprovodnikov* [Ion doping of semiconductors]. Moscow. World. 1973. (in Russian)

106.Zorin E.I., Pavlov P.V., Tetel'baum D.I. *Ionnoe legirovanie poluprovodnikov* [Ion doping of semiconductors]. Moscow. Energiia. 1975. (in Russian)

107.*Tekhnologiia ionnogo legirovaniia* [Ion doping technology] Moscow. Soviet radio. 1974, p. 158. (in Russian)

108.Gibbons J. E. ION Implantation in Semiconductors. Part 2. Damage Production and Annealing. Proc. IEEE, 1972, Vol. 60, Iss 9, p. 1062 – 1096.

109.Chadderton L.T. Eisen F. H. On the annealing of Damage Produced by B^+ Ion Implantation of Si Single Crystals. In: Ion Implantation. Ed. F. H., Eisen, L. T. Chadderton. London, Gordon and Breach. Sci. Publ., 1971, p. 445—454.

110.Chadderton L. T., Eisen F. H. On the annealing of Damage Produced by B^+ Ion Implantation of Si. Single Crystals. Rad. EFF., 1971, Vol.7, Iss 1-2, p. 129.

111. Grilhe j., Seshan K., Washburn J. On the Possibility of Nucleating Loops with Burgers Vectors DC by the Clustering of Interstitials.-

Rad. Eff., 1975, Vol. 25, Iss 1-2, p. 115-118,

112.Maksimov P.K., Luk'ianchuk T.I., Piskunov L.I. *Dva mekhanizma formirovaniia defektnoi struktury v kremnii pri ionnom vnedrenii primesi i posleduiushchem otzhige* [Two mechanisms of the formation of a defective structure in silicon upon ion implantation of an impurity and subsequent annealing]. Reports of the USSR Academy of Sciences. Vol. 223, Iss. 6, p. 1351-1354. (in Russian)

113.Baranov A.I., Gerasimenko N.N., Dvurechenskii A.V., Smirnov L. P. *Otzhig krupnykh vakansionnykh klasterov* [Annealing large vacancy clusters]. Semiconductor Physics and Technology. 1974. Vol.11, Iss. 1, p. 92 - 99. (in Russian)

114.Baranov A.I. *Otzhig slozhnykh defektov v obluchennykh poluprovodnikakh* [Annealing of complex defects in irradiated semiconductors]. In book: *Radiatsionnye effekty v poluprovodnikakh* [Radiation effects in semiconductors]. Novosibirsk. Science. 1979. (in Russian)

115.Gerasimenko N.N., Dvurechenskii A.V., Kachurin G.A. and other. *Radiatsionnyi otzhig defektov, obrazuiushchikhsia pri bombardirovke kristallov ionami* [Radiation annealing of defects formed during the bombardment of crystals by ions]. Semiconductor Physics and Technology. 1972, Vol. 6, Iss. 9, p. 1834 – 1835. (in Russian)

116.Mordkovich B.N., Solov'ev P.P., Temper E.M., Kharchenko V.A. *O radiatsionnom otzhige defektov v obluchennom neitronami kremnii* [On radiation annealing of defects in neutron-irradiated silicon]. Semiconductor Physics and Technology. 1974, Vol. 8, Iss. 5, p. 1024 – 1025. (in Russian)

117.Vikhrev B.N., Gerasimenko N.N., Dvurechenskii A.V., Smirnov L.P. *Vzaimodeistvie v kremnii atomov vodoroda s defektami, vvedennymi ionnoi bombardirovkoi* [Interaction of hydrogen atoms in silicon with defects introduced by ion bombardment]. Semiconductor Physics and Technology. 1974, Vol. 8, Iss.7, p. 1345 – 1348. (in Russian)

118.Stein H.G. Bonding and Thermal Stability of implanted Hydrogen in Silicon. Electron. Mat., 1975, Vol. 4, Iss. 1, p. 159 – 174.

119.Picraux S.T., Vook F.L. Ionization, Thermal and Flux Dependences of Implantation Disorder in Silicon. Rad., Eff., 1971, Vol. 11, Iss 2, p.

179 – 192.

120.Kachurin G.A., Pridachin H.B., Smirnov L.P. *Otzhig radiatsionnykh defektov impul'snym lazernym izlucheniem* [Annealing of radiation defects by pulsed laser radiation]. Semiconductor Physics and Technology. 1975, Vol. 9, Iss. 7, p. 1428-1429. (in Russian)

121.Antonenko A.X., Gerasimenko N.N., Dvurechenskii A.V. and other *Raspredelenie vnedrennoi v kremnii primesi posle lazernogo otzhiga* [Distribution of impurities embedded in silicon after laser annealing]. Semiconductor Physics and Technology. 1976, Vol. 40, Iss. 1, p. 139 - 141. (in Russian)

122.Khaibullin I.B., Shtyrkov E.I., Zaripov M.M. and other. *Otzhig ionno legirovannykh sloev pod deistviem lazernogo izlucheniia* [Annealing of ion-doped layers under the action of laser radiation.] Moscow VINITI, dep. № 2661-74, p. 32. (in Russian)

123.Pfister J.P. Radiation Enhanced Diffusion in Si. - In: 7-th Intemat. Conf. on the Physics of Semiconductors, Paris - Royanmaent, 1964, Vol. 3, p. 281—285.

124.Brelot A. Enhanced Diffusion in Ge. In: Radiation Effects in Semiconductors. N. Y., Plenum Press, 1968, p. 460-465.

125.Tsuchimoto T., Tokuyama T. Enchanced Diffusion of Boron and Phos phorus in Si during Hot Substrate Ion Implantation. Rad. Eff., 1970, Vol. 6, p.121-129.

126.Ohmura Y., Mimura S., Kanazawa M. e. a. Enhanced Diffusion and Dose Rate Dependence of Sb and P in Si by Proton Irradiation. Rad. Eff., 1972, Vol, 15, Iss 2, p.167-174.

127.Miner B. L., Nelson D.G., Gibbson J.F. Enchanced Diffusion in Si and Ge by Linght Ion Implantation. J. Appl. Phys., 1972, Vol.43, p.3468 – 3480.

128.*Radiatsionno – aktiviruemye protsessy v kremnii* [Radiation-activated processes in silicon]. Tashkent. Fan Publishing House. 1977, 165 p. (in Russian)

129.Seriapin V.G., Seriapina N.V., Smirnov L.P. and other. *Pererarspredelenie bora v kremnii pri obluchenii bystrymi elektronami* [Redistribution of boron in silicon upon irradiation with fast electrons]. Semiconductor Physics and Technology. 1973, Vol. 7, Iss. 1, p. 183 - 185. (in Russian)

130.Gerasimenko N.N., Obodnikov V.I., Smirnov L.P., Sokolov P.A. *Diffuziia fosfora v kremnii pod deistviem oblucheniia ionami vodoroda pri povyshennoi temperature* [Diffusion of phosphorus in silicon by irradiation with hydrogen ions at elevated temperature]. Semiconductor Physics and Technology. 1975, Vol. 9, Iss. 11, p. 2220 – 2221. (in Russian)
131.Stas' V.F., Smirnov L.P. *Radiatsionno – uskorennaia diffuziia fosfora v germanii* [Radiation-accelerated diffusion of phosphorus in germanium]. Semiconductor Physics and Technology. 1979, Vol. 13, Iss. 4. (in Russian)
132.Abe T., Ohmura Y., Konaka M. e.a. Radiation Enhanced Diffusion and Its Application to Devise Fabrication. Proc. 2nd Conf. Solid State Devices, Tokyo, 1970, p. 16 -23.
133.Pigg J. C., Grawford J.H. Radiation Effects and Their Annealing in Co^{60} Gamma Irradiated Sb- Doped Germanium. Phys. Rev., 1964, Vol.135, p. A1141 – A1150.
134.Vitovskii N.A., Maksimov M., Mashovets T.V. *Kinetika nakopleniia radiatsionnykh defektov v germanii vysokoi chistoty pri γ-obluchenii* [Kinetics of the accumulation of radiation defects in high purity germanium under γ-radiation]. Semiconductor Physics and Technology. 1970, Vol. 4, Iss. 12, p. 2276-2284. (in Russian)
135.Emtsev V.V., Mashovets T.V. *O kompleksakh vakansiia – donor v germanii* [About vacancy complexes - donor in germanium]. Pis'ma v ZhETF. 1971, Vol. 13, Iss. 12, p. 675—679. (in Russian)
136.Abdulaev A., Emtsev V.V., Korchazhkina R.L., Mashovets T.V. *Protsess obrazovaniia γ-radiatsionnykh defektov v germanii, legirovannom fosforom* [The process of formation of γ-radiation defects in germanium doped with phosphorus]. Semiconductor Physics and Technology. 1971, Vol. 5, Iss. 11, p. 2229. (in Russian)
137.Vitovskii N.A., Emtsev V.V., Kotina I.M., Mashovets T.V. *Ob odnoi zakonomernosti radiatsionnogo defektoobrazovaniia v germanii* [About one regularity of radiation defect formation in germanium]. Semiconductor Physics and Technology. 1971, Vol. 5, Iss. 11, p. 2231. (in Russian)
138.Emtsev V. V., Korkhashkina R. L., Mashovets VOL. V. Vacancy - Donor Complexes in Irradiated Phosphorus-Doped Germanium. Phys.

Stat. Sol, (a), 1972, Vol.10, Iss 1, p. 43—48.

139.Tkachev V.D., Urenev V.I. *Vzaimodeistvie elementarnykh defektov reshetki s atomami donornoi primesi v germanii* [Interaction of elementary lattice defects with atoms of a donor impurity in germanium]. Semiconductor Physics and Technology. 1971, Vol. 5, Iss. 8, p. 1516—1521. (in Russian)

140.Basman A.R., Gerasimov A.B., Gogotishvili M.K. and other. *Vliianie donornykh primesei na kinetiku otzhiga radiatsionnykh defektov v Ge* [The effect of donor impurities on the kinetics of annealing of radiation defects in Ge]. Semiconductor Physics and Technology. 1975, Vol. 7, Iss. 7, p. 1377 - 1381. (in Russian)

141.Emtsev V.V., Mashovets T.V., Ryvkin P.M. *Protsess defektoobrazovaniia v beskislorodnom germanii pri gamma-obluchenii* [The process of defect formation in oxygen-free germanium under gamma radiation]. In book: *Radiatsionnye defekty v poluprovodnikakh* [Radiation defects in semiconductors]. Minsk, Publishing House of the Belarusian State University named after V.I. Leninю 1972, p. 122—124. (in Russian)

142.Mashovets T.V., Emtsev V.V., Abdurakhmanova P.N. *Model' protsessa obrazovaniia γ-radiatsionnykh defektov v Ge, legirovannom primesiami V gruppy* [Model of the process of formation of γ-radiation defects in Ge doped with group V impurities]. Semiconductor Physics and Technology. 1974, Vol.8, Iss.1. p. 96-104 (in Russian)

143.Emtsev V.V., Klinker M.I., Mashovets T.V. *O vozmozhnom mekhanizme rasseianiia nositelei zariada v Ge s tochechnymi defektami* [On a possible mechanism for scattering of charge carriers in Ge with point defects]. Pis'ma v ZhETF. 1974, Vol.19, Iss.9, p.575 – 599. (in Russian)

144.Smirnov L.P., Stas' V.F., Khainovskaia V.V. *Opredelenie stepeni kompensatsii germaniia* [Determination of the degree of compensation for germanium]. Semiconductor Physics and Technology. 1971, Vol.5, Iss.8, p. 1849 – 1850. (in Russian)

145.Astakhov V.M., Golobokov Iu.N., Stas' V.F., Smirnov L.P. Otzhig germaniia i kremniia, obluchennogo elektronami [Annealing of germanium and silicon irradiated with electrons]. Semiconductor Physics and Technology. 1973, Vol.7, Iss. 5, p. 1414 – 1417. (in

Russian)
146. Mashovets T.V. *Tochechnye defekty v almazopodobnykh poluprovodnikakh* [Point defects in diamond-like semiconductors]. Abstract of a doctoral dissertation. St. Petersburg. 1975. (in Russian)
147. Hasiguti R.R., Tanaka K., Takahashi S. ESR Studies of Electron Irradiation Damage and Its Annealing in As- Doped *n*-Type Germanium. In: Radiation Effects in Semiconductors. N.Y., Plenum Press, 1968, p.89-98.
148. Whan R.E. Evidence for Low-Temperature Motion of Vacancies in Ge. –Appl. Phys. Lett., 1965, Vol. 6, p. 221.
149. Whan R.E. Investigations of Oxygen – Defects Interactions Between 25 and 700K in Irradiated Germanium. Phys. Rev., 1965, v.140, p. A690 - A698.
150. Baldwin J.A. Electron Paramagnetic Resonance in Irradiated Oxygen- Doped Germanium. J. Appl. Phys., 1965, Vol 36, Iss 3, part. 1, p. 793 – 795.
151. Hiraki A., Cleland J.W., Crawford J. H. Experimental Evidence for Production and Annihilation of Interstitial Impurities in Electron Irradiated *n*- Type Germanium. In: Radiation Damage in Semiconductors. N.Y., Plenum Press, 1968, p.224 - 231.
152. Hiraki A., Cleland J.W., Crawford J. H. Observation of Irradiation Induced Interstitial Copper Impuritiy in Germanium. J. Appl. Phys., 1967, Vol. 38, Iss 9, p. 3519 – 3527.
153. Boiarkina N.I., Smirnov L.P., Stas' V.F. *Obluchenie elektronami i otzhig germaniia r- tipa, legirovannogo zolotom* [Electron irradiation and annealing of p-type germanium doped with gold]. Semiconductor Physics and Technology. 1975, Vol.9, Iss.2, p. 376 – 378. (in Russian)
154. Boiarkina N.I., Smirnov L.P., Stas' V.F. *Umen'shenie kontsentratsii rekombinatsionnykh tsentrov, sviazannykh s zolotom, pri obluchenii germaniia elektronami* [A decrease in the concentration of recombination centers associated with gold upon irradiation of germanium with electrons]. Semiconductor Physics and Technology. 1974, Vol.8, Iss.11, p. 2201 – 2202. (in Russian)
155. Mamontov F.M., Baryshev N.P. *Vliianie elektronnogo oblucheniia na elektricheskie svoistva germaniia, legirovannogo tsinkom i rtut'iu*

[The effect of electron irradiation on the electrical properties of germanium doped with zinc and mercury]. Semiconductor Physics and Technology. 1976, Vol.10, Iss.8, p. 1483 – 1485. (in Russian)

156. Sinchuk I.K., Tkachev V.D., Urenev V.I. *Vzaimodeistvie litiia s radiatsionnymi defektami v chistom n-germanii* [Interaction of lithium with radiation defects in pure n-germanium]. Semiconductor Physics and Technology. 1971, Vol.5, Iss.11, p. 2195 – 2197. (in Russian)

157. Kasherininov P.G., Matveev O.A., Tomasov A.A. *Obrazovanie assotsiatsii melkii donor – glubokii aktseptor v germanii* [Association formation shallow donor - deep acceptor in germanium]. Semiconductor Physics and Technology. 1974, Vol.8, Iss.4, p. 799 – 800. (in Russian)

158. Konopleva R.F., Novikov P.R. *Elektricheskie svoistva germaniia, obluchennogo neitronami pri 77K* [Electrical properties of germanium irradiated with neutrons at 77K]. Solid state physics, 1964, Vol.6, Iss.4, p. 1062 – 1065. (in Russian)

159. Gerasimov A.B., Konovalenko B.N. *O nizkotemperaturnom obluchenii germaniia bystrymi elektronami* [On low-temperature irradiation of germanium with fast electrons]. Semiconductor Physics and Technology. 1964, Vol.6, Iss.10, p. 3184 – 3186. (in Russian)

160. Gerasimov A.B., Dolidze N.D. Kakhidze N.G. and other. *Vliianie sveta na nizkotemperaturnyi otzhig radiatsionnykh defektov v germanii* [The effect of light on low-temperature annealing of radiation defects in germanium]. Semiconductor Physics and Technology. 1967, Vol.1, Iss.7, p. 982 – 985. (in Russian)

161. Stas' V.F., Smirnov L.P. *Nizkotemperaturnoe obluchenie i otzhig germaniia p- tipa* [Low-temperature irradiation and annealing of p-type germanium.]. Semiconductor Physics and Technology. 1968, Vol.2, Iss.9, p. 1369 – 1371. (in Russian)

162. Trueblood D. L. Electron Paramagnetig Resonance in Electron Irradiated Germanium. Phys. Rev., 1967, Vol.161, p. 828 – 833.

163. Stas' V.F., Smirnov L.P. *Obluchenie i otzhig germaniia p- tipa* [Irradiation and annealing of p-type germanium.]. Semiconductor Physics and Technology. 1970, Vol.4, Iss.2, p. 276 – 281. (in

Russian)
164.Flanagan T. M., Klotz E. E. Bombardment – Produced Defects in p-Type Germanium at Low Temperatures. Phys. Rev., 1968, Vol.167, Iss 3, p. 789 – 800.
165.Shimotomai M., Hasiguti R. R. The 220K Defect in Electron Irradiated *p*- Type Germanium. Rad. Eff., 1971, Vol.9, Iss 1-2, p. 47 – 49.
166.Saito H., Fukuoka N., Tatsumi J. Annealing of Point Defects in *p*-Type Germanium After Electron Irradiationat Liquit Nitrogen Temperature. Rad. Eff., 1971, Vol.8, Iss 3-4, p. 171 -175.
167.Lang D.V. Review of Radiation – Induced Defects in III-V Compaunds. In: Radiation Effects in Semiconductors. Conf. Ser. N 31, Inst. of Phys., Bristol – London, 1976, p. 70 -94.
168.Mashovets T.V. *Sostoianie i perspektivy radiatsionnoi fiziki poluprovodnikov gruppy* A^3V^5 [State and prospects of radiation physics of A^3B^5 semiconductors]. In book: *Radiatsionnaia fizika nemetallicheskikh kristallov* [Radiation Physics of Non-Metallic Crystals]. Kiev. Naukova dumka. 1971. p. 5 – 27. (in Russian)
169.Galushka A.P., Konozenko I.D. *Sostoianie i perspektivy radiatsionnoi fiziki poluprovodnikov gruppy* A^2V^6 [State and prospects of radiation physics of A2B6 semiconductors]. In book: *Radiatsionnaia fizika nemetallicheskikh kristallov* [Radiation Physics of Non-Metallic Crystals]. Kiev. Naukova dumka. 1971. p. 28 – 41. (in Russian)
170.Watkins G. D. Lattice Defects in II-VI Compounds. In: Radiation Effects in Semiconductors. Conf. Ser. N 31. Inst. of Phys. Bristol – London. 1976. p. 95 – 111.
171.Eisen F. H. Stage- II Recovere in Electron Irradiated InSb. Phys. Rev. 1966. Vol. 148, Iss. 2, p. 828 – 838.
172.Eisen F. H. Recovere of Electron Radiation Damage in *n*-Type InSb. Phys. Rev. 1961. Vol. 123, Iss 3, p. 736 – 744.
173.Eisen F. H. Orientation Dependence of Electron Radiation Damage in InSb. Phys. Rev. 1964. Vol. 135, Iss 5A. p. A1394 – A1399.
174.Eisen F. H., Bickel P. W. Electron Damage Threshold in InSb. Phys. Rev. 1959. Vol. 115, Iss 2, p. 345 – 346.
175.Aukerman L. W. Electron Irradiation of Indium Antimonide. - Phys.

Rev., 1959. Vol. 115, Iss 5, p. 1125 – 1132.
176.Mashovets T.V., Khansevarov R.Iu. *Nizkotemperaturnoe obluchenie i otzhig sur'mianistogo indiia* [Low-temperature irradiation and annealing of antimony indium]. Semiconductor Physics and Technology. 1966. Vol.8, p.1690–1697. (in Russian)
177.Mashovets T.V., Khansevarov R.Iu. *Energeticheskii spektr defektov, obrazuiushchikhsia v antimonide indiia pri nizkotemperaturnom γ-obluchenii* [Energy spectrum of defects formed in indium antimonide under low-temperature γ-radiation]. Solid state physics. 1965. Vol.7, Iss. 7, p.2229 –2231. (in Russian)
178.Vitovskii N.A., Mashovets T.V., Khansevarov R.Iu., Chelustka B. *Issledovanie izokhronnogo i izotermicheskogo otzhiga InSb, obluchennogo rentgenovskimi i γ- kvantami* [Investigation of isochronous and isothermal annealing of InSb irradiated with x-ray and γ-quanta]. Semiconductor Physics and Technology. 1967, Vol.1, Iss.5, p. 766 – 773. (in Russian)
179.Abdullaev A., Vitovskii N.A., Mashovets T.V., Mustafakulov D. *Protsess oblucheniia defektov struktury v antimonide indiia pri γ-obluchenii* [The process of irradiation of structural defects in indium antimonide during γ-irradiation]. Semiconductor Physics and Technology. 1975, Vol.9, Iss.2, p. 282 – 286. (in Russian)
180.Chelustka B., Khansevarov R.Iu., Mashovets T.V., Kozlova I. R. *Obrazovanie strukturnykh defektov pri nizkotemperaturnom rentgenovskom obluchenii sur'mianistogo indiia* [The formation of structural defects during low-temperature x-ray irradiation of antimony indium]. Solid state physics. 1967, Vol.9, Iss. 1, p.338–340. (in Russian)
181.Arnold G. W., Vook F. L. Production of Defects in InSb by X- Rays. Phys. Rev., 1965, Vol.137, Iss 6A, p. A1839 – A1842.
182.Vitovskii N.A., Vikhlii G.A., Mashovets T.V. *Radiatsionno-stimulirovannoe kompleksoobrazovanie v antimonide indiia r-tipa* [Radiation-stimulated complexation in p-type indium antimonide]. Semiconductor Physics and Technology. 1972. Vol.6, Iss.10, p. 1995 – 2002. (in Russian)
183.Abdullaev A., Vitovskii N.A., Mashovets T.V., Morozov Iu. G. *Ionizatsionnyi mekhanizm sozdaniia defektov struktury v antimonide*

indiia [Ionization mechanism for creating structural defects in indium antimonide]. Semiconductor Physics and Technology. 1975, Vol.9, Iss.1, p. 68 – 75. (in Russian)

184. Vitovskii N.A., Vikhlii G.A., Galavanov V.V. and other. *Obrazovanie radiatsionnykh defektov v antimonide indiia pri doporogovykh energiiakh izlucheniia* [The formation of radiation defects in indium antimonide at subthreshold radiation energies]. Semiconductor Physics and Technology. 1969, Vol.3, Iss.1, p. 132 – 134. (in Russian)
185. Cleland J.W., Crawford J. H. Neutron Irradiation of Indium Antimonide. Phys. Rev., 1954, v.95, N 5, p. 1177 – 1182.
186. Vodop'ianov L.K., Kurdiani N.I. *Elektricheskie svoistva sur'mianistogo indiia, obluchennogo neitronami pri 77K i elektronami pri 300K* [Electrical properties of antimony indium irradiated with neutrons at 77K and electrons at 300K]. Solid state physics. 1965, Vol.7, Iss. 9, p. 2749 – 2753. (in Russian)
187. Vodop'ianov L.K., Kurdiani N.I. *O razuporiadochennykh oblastiakh v InSb, obuslovlennykh oblucheniem bystrymi neitronami* [On disordered regions in InSb due to fast neutron irradiation]. Semiconductor Physics and Technology. 1967, Vol.1, Iss.5, p. 646 – 648. (in Russian)
188. Kaniashvili R.G., Kurdiani N.I. *Vliianie neitronnogo oblucheniia na podvizhnost' elektronov v plasticheski deformirovannykh kristallakh* [The effect of neutron irradiation on the mobility of electrons in plastically deformed crystals]. Semiconductor Physics and Technology. 1971, Vol.5, Iss.6, p. 1170 – 1173. (in Russian)
189. Foyt A.G., Lindley W.T., Donnelly J.P. *n–p* Junction Photodetectors in InSb Fabricated by Proton Bombardment. Appl. Phys. Lett., 1970, Vol.16, Iss 9, p. 335 – 337.
190. Vavilov V.P., Vodop'ianov L.K., Kurdiani N.I. *Deistvie oblucheniia bystrymi neitronami i elektronami na elektricheskie svoistva sur'mianistogo indiia* [The effect of irradiation with fast neutrons and electrons on the electrical properties of antimony indium]. In book: *Radiatsionnaia fizika nemetallicheskikh kristallov* [Radiation physics of non-metallic crystals]. Kiev. Naukova dumka. 1967, p. 191 – 200. (in Russian)

191. Vitovskii N.A., Mashovets T.V., Oganesian O.V. *Vliianie intensivnosti elektronnogo oblucheniia na defektoobrazovanie v antimonide indiia* [Influence of the intensity of electron irradiation on defect formation in indium antimonide]. Semiconductor Physics and Technology. 1978. Vol.12, Iss.11, p. 2143 – 2148. (in Russian)

Chapter 4 Silicon Nuclear Doping Technology

In chapter 2 and 3, the physical foundations of the processes occurring in semiconductor materials upon irradiation with various nuclear particles were considered. The possible nuclear reactions and the expected distribution of the formed impurities over the volume of doped crystals are analyzed. The prospects of using radiation technology for doping monocrystalline silicon with phosphorus using nuclear reactions using slow neutrons to produce *n*-type material and aluminum using photo-nuclear reactions under the influence of bremsstrahlung γ radiation to produce *p*-type material have been noted. However, photonuclear reactions have not yet received real practical application. Therefore, in this chapter we consider only the technology of doping silicon with the help of slow neutrons, because this method can be used to obtain *p*-type material. In particular, it is shown how the physical principles of nuclear doping are implemented in practice, what technical and economic indicators are achieved and what equipment is required for this.

4.1. The technological scheme of the process of nuclear doping

A diagram of the nuclear doping process based on the accumulated experience at the WWR-t reactor and the available data (see, for example, [1,2]) is shown in Fig. 4.1. The main operation, which determines the main qualitative and economic indicators of the process, is the irradiation of silicon ingots with slow neutrons.

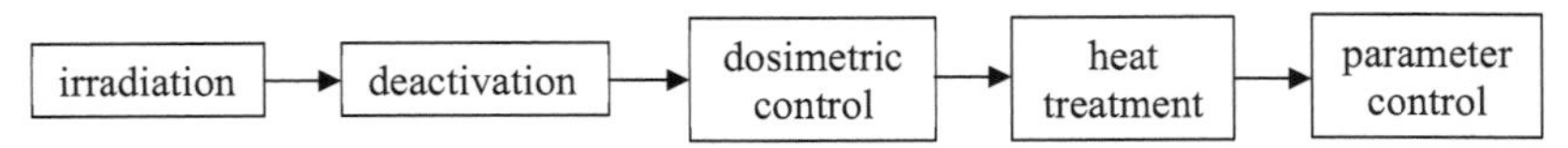

Figure 4.1. The technological scheme of nuclear doping.

Annealing operation is equally important. It is required to remove radiation defects and stabilize the properties of nuclear-doped silicon (NDS) after irradiation.

The decontamination operation is caused by the need to clean the ingots from surface radioactive contamination. For the correct choice of means of protection against radioactive radiation and monitoring the effectiveness of decontamination, dosimetric control is introduced. The need for such operations as preparation of ingots for irradiation and their removal after doping is obvious.

The specificity of technological operations lies in the very principle of doping. As a result of irradiation with slow neutrons, along with the target impurities inside the volume and on the surface of the ingots, radioactive atoms of side impurities are formed. Induced radioactivity also appears in structural materials. Therefore, in the practical implementation of nuclear doping technology, direct contact of the personnel with radioactive samples and devices should be excluded whenever possible. In addition, measures must be taken to prevent the spread of radioactivity beyond the limits of special rooms.

Due to the fact that the electrophysical properties and their homogeneity in single crystals of NDS depend on a number of factors that are not typical for metallurgical methods of introducing impurities, specific requirements arise for the process of obtaining the initial single crystal silicon. Therefore, knowledge of all these factors is very important, because the final result of doping will depend on the quality of the starting material, the parameters of the radiation source, as well as the choice of optimal irradiation modes and the subsequent heat treatment of irradiated ingots.

4.2. Ensuring uniform irradiation of ingots in nuclear reactors

As noted in Chapter 2, radioactive sources, charged particle accelerators and nuclear reactors can be used as neutron sources for nuclear doping. However, for practical purposes of doping, the most suitable sources of neutron intensity and energy are nuclear reactors based on the use of fission of uranium nuclei. The design feature of nuclear reactors is the presence of an active zone in the form of a compact assembly of fuel cells containing a natural mixture of uranium or uranium isotopes enriched in the U^{235} isotope, as well as a moderator, which, depending on the type of reactor, is most often used in plain or heavy water and graphite (see, for example, [3]).

The energy spectrum of neutrons arising from the fission of U^{235} is shown in fig. 4.2. Simultaneously with the fission neutrons in the reactor at each instant of time there is a large number of previously formed neutrons that are at different stages of deceleration, up to thermal equilibrium with the moderator medium. In the region where the uranium rods are located, an approximate equality of thermal and fast neutrons of the fission spectrum is observed. As you move away from the rods, the neutron intensity of the fission spectrum decreases exponentially, but the relative fraction of neutrons with lower energy increases. Ultimately, it is possible to obtain neutrons in thermal equilibrium with the moderator medium and obeying the Maxwellian distribution. However, this can only be achieved in special devices placed in the reactor near the core in called thermal columns, with a corresponding loss in neutron flux density.

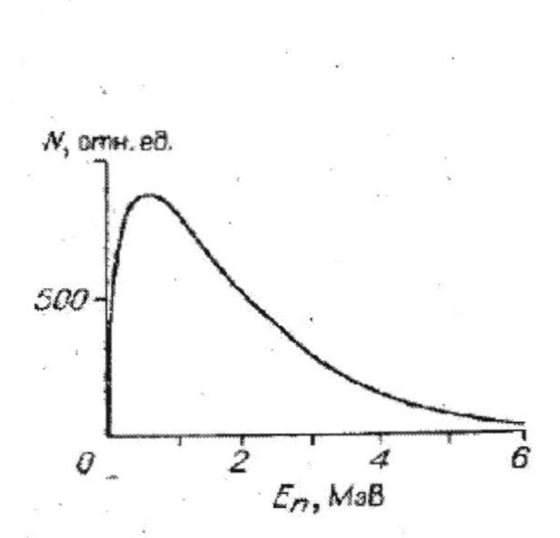

Figure 4.2. Energy distribution of neutrons in fission U^{235}.

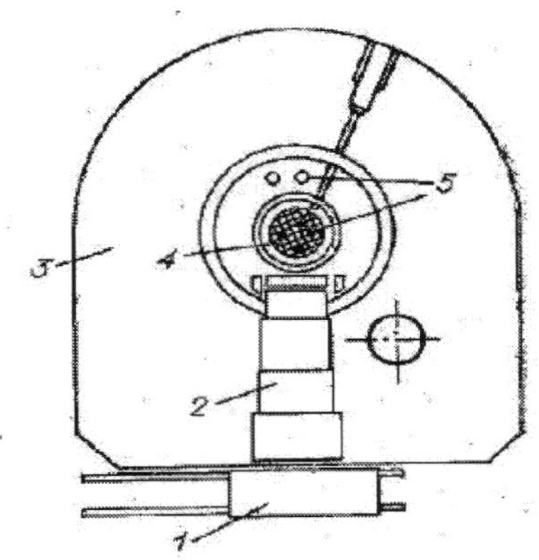

Figure 4.3. Cross section of the WWR-t reactor [3].
1-protection of the niche of the heat column; 2-niche of the heat column; 3 - reactor protection; 4-active zone; 5-channels for irradiation.

Usually, each specific reactor is characterized by a certain ratio of the number of slow neutrons that are of interest for nuclear doping to the number of fast neutrons that create mainly radiation defects. Depending on the design of the core and those or other devices, in particular, channels for various purposes, the indicated ratio can vary within significant limits and within one reactor. Such, for example, is the situation in the WWR-t reactor, shown in fig. 4.3 [3]. In the general case, the greater the ratio of slow to fast neutron flux, the softer the neutron spectrum and the more favorable the conditions for nuclear doping.

Another factor important for nuclear doping is the degree of constancy of the neutron flux density along the length and radius of the core, since the possibility of uniform irradiation of sufficiently large ingots depends on this.

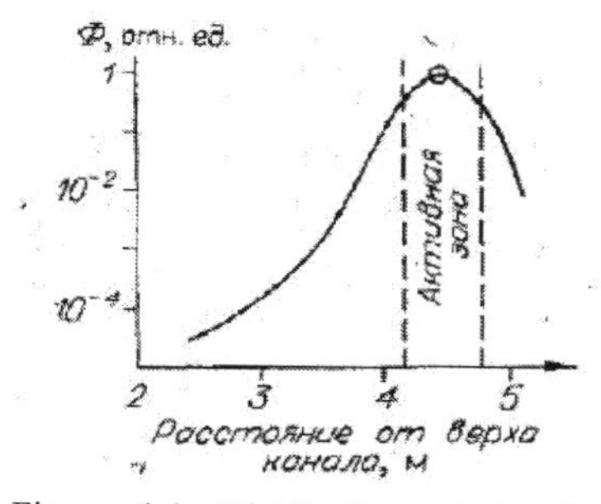

Figure 4.4. Distribution of the flux density of slow neutrons over the height of the channel in the WWR-t reactor. The circle marks the site of irradiation of silicon ingots.

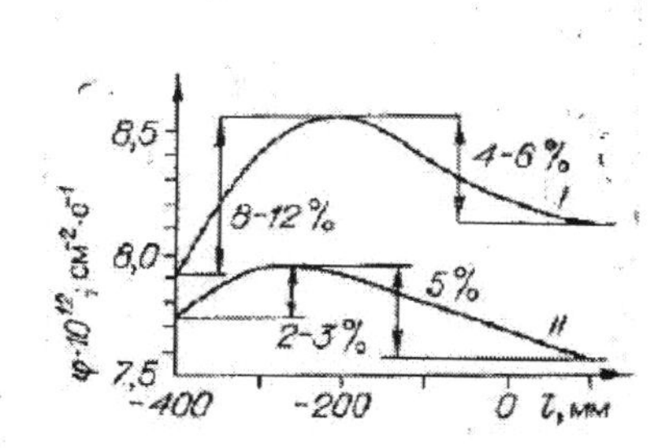

Fig. 4.5. The usual (*I*) and modified (*II*) distribution of the neutron flux density is within 0.5 m along the channel height of the heavy water reactors DIDO and PLUTO [9].

It turns out that reactors with a moderator of heavy water have advantages over ordinary water reactors both in terms of the relative fraction of slow neutrons in the neutron spectrum and in terms of the nature of the distribution of neutron flux density along the length of the core (Fig. 4.4, 4.5). It can be seen that with an accuracy of 5-10% in a heavy-water reactor, the neutron flux density can be considered constant over a length of about 500 mm [4], and in the case of a light-water reactor, over a length of only 200-250 mm (see Fig. 4.4 and operation [5]).

In the radial direction, the neutron flux density varies insignificantly within the active zone, and decreases exponentially outside it (Fig. 4.6). It is possible to reduce or compensate for the inhomogeneity of the neutron flux in the irradiation zone of the ingots using special techniques (see below). One of these techniques, which can be arbitrarily called passive, is reduced to smoothing out the inhomogeneity using a special screen or channel shape that has such a profile so that neutrons are absorbed more in those areas where the flux density is higher. The effectiveness of this technique is also shown in fig. 4.5.

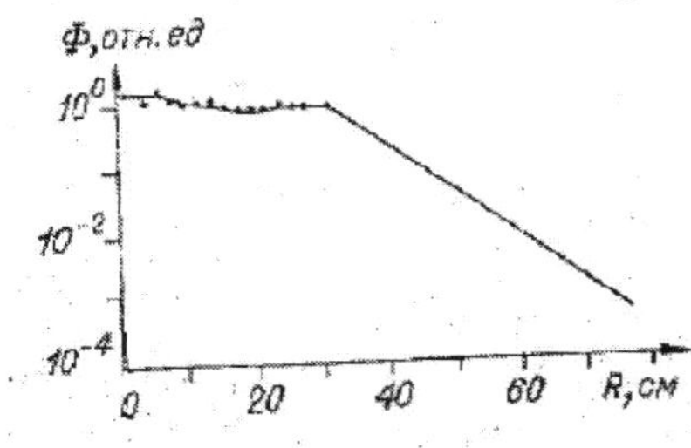

Figure 4.6. Distribution of neutron flux density along the radius of the WWR-t reactor

Depending on the design and purpose, specific reactors usually provide for the possibility of irradiating materials in various places of the working volume, differing both in intensity and composition of radiation (contribution of γ radiation, ratio of slow and fast neutrons). This is achieved using vertical or horizontal channels that can be located inside the core or outside it, for example, in a tank with a moderator surrounding the core (see Fig. 4.3). In the absence of suitable channels, irradiation can be carried out in specially designed autonomous devices introduced into the reactor at certain places, for example, inside a graphite reflector, as is done in the reactor of the University of Missouri [5].

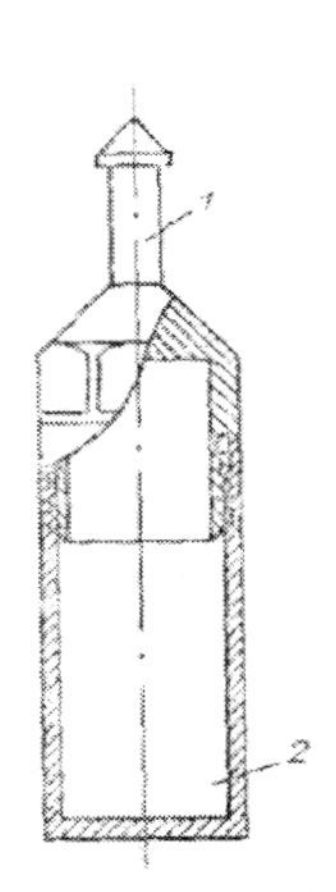

Figure 4.7. Ampoule for transporting silicon ingots. 1 - ampoule head; 2 - a glass.

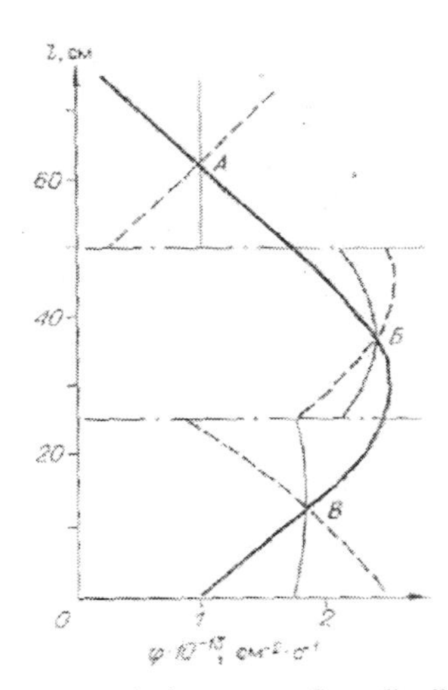

Figure 4.8. Typical neutron flux distribution profile over the height of the reactor working zone (thick solid line). The dotted line is a mirror image of the distribution sites at the location of the three silicon ingots, thin lines are the total flow distribution along the length of the ingots after irradiation in the 180° rotation mode [5].

For ease of transportation, protection from mechanical damage, and in some cases to create a local atmosphere, to provide local heating or cooling conditions, samples and ingots are irradiated in special ampoules, the material of which must have the necessary mechanical strength, thermal and radiation resistance, and low induced radio activity. Most often, pure aluminum and some types of plastics, such as polystyrene, are used as such materials [4]. On the fig. 4.7 shows one of the designs of a sealed ampoule for irradiating various materials, including silicon ingots.

In order to prevent undesirable excessive heating of the samples during irradiation, mainly due to the γ-component of the radiation from the reactor, gas is pumped through the channel with ampoules or directly through the holes in the ampoules, for example air, or coolant, which can be played by ordinary or heavy water. But even with intensive cooling by heavy water flow, the temperature of silicon ingots during irradiation reaches 70°C [4].

The original way to prevent the heating of silicon ingots was implemented in the WWR-t reactor [6]. The through channel, in which silicon is irradiated, is equipped with special protection from the core, made of a material that effectively absorbs gamma radiation and is transparent to thermal neutrons, which significantly reduced the temperature of irradiated ingots. The protection is equipped with an autonomous cooling system with running water (Fig. 4.9).

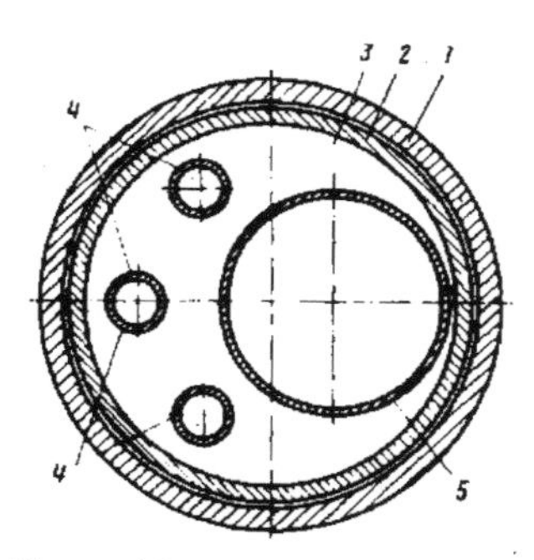

Figure 4.9. Design of a through technological channel with a screen (WWR-t reactor).
1 - external channel; 2 - technological channel; 3 - protection from lead (screen); 4 - assembly cooling system with flowing water; 5 - working channel

Ensuring conditions for uniform irradiation of extended ingots, especially with a large ratio of length to diameter, is at the present stage of nuclear doping technology an independent scientific and technical task for each specific reactor. It is likely that in the near future there will appear reactors specially designed for nuclear doping, but for now it is necessary to adapt the existing ones to the new technology.

As already noted, the radial non-uniformity of the neutron flux over the diameter of real silicon ingots up to 200 mm in diameter is not very large and is practically eliminated by rotating the ingot during irradiation. In this case, forced rotation can be achieved using both conventional mechanisms and a special ampoule shape placed in a cooling D_2O stream [4].

The main problem is related to ensuring uniform exposure along the length of the ingot. It was indicated above that within certain limits this can be achieved by placing the ingot in such a section of the reactor channel where the neutron flux density can be considered constant (see Fig. 4.4 and 4.5). However, with such a static doping method, only a small part of the working channel length is used, which negatively affects the performance of the process.

This difficulty can be partially circumvented by using the technique used in [5] and shown schematically in Fig. 4.8. Let three ingots be arranged vertically with centers of gravity at points A, B, C, then the distribution profile of dopants in each ingot should repeat the corresponding section of the neutron distribution profile. If under such conditions only half of the estimated time is irradiated with the ingots and then "tilted" by 180°, then in the remaining time the lower part of each ingot is irradiated similarly to the upper one, and vice versa. It turns out the same as if stationary ingots in the remaining time were irradiated with a neutron flux, the profile of which is a mirror image of the original profile in the vertical plane. The total doping result along the length of the ingots in this mode is determined by the sum of the initial and mirror sections of the profile and is generally much more uniform, and in the straight sections of the profile (point A) - strictly uniform.

Two variants of the technological scheme of nuclear doping under static irradiation are shown in Fig. 4.10. In option a), carried out at the WWR-t reactor in 1974, ampoules from the pre-loading chamber are placed by the capture mechanism in the working channel. After irradiation, the ampoules with the same device are removed from the channel and transferred to the transport channel, through which they fall into the "hot" chamber, designed to work with radioactive materials. Here, the ingots are remotely removed from the ampoules, kept for several days to reduce radioactivity and then sent for decontamination. After decontamination, work with irradiated ingots is no longer dangerous and is carried out in ordinary rooms. All samples after extraction from ampoules and decontamination are subjected to mandatory dosimetric control.

A similar irradiation scheme was described in [6, 7]. The only difference is that in the working channel there is a special movable support on which the container is mounted. The design features of the channel and container are shown schematically in Fig. 4.11. A special hole is provided in the bottom of the container, into which the pin of the mechanism of rotation of the container during irradiation enters. After loading silicon, the container moves to the working area, where it is fixed with a grid, the position of which relative to the center of the active zone can be changed.

In the second version of the static irradiation regime described in [4] and shown in Fig. 4.10 b), ampoules with non-irradiated samples are first loaded into the glove box, from where they already automatically fall onto the conveyor. Structurally, the conveyor is a 100 interlocked bogies in the form of a closed loop, which is driven by an electric motor. Each trolley is designed for the installation of one ampoule with silicon. Carts are consequently loaded with ampoules as they are released after each loading cycle and subsequent unloading.

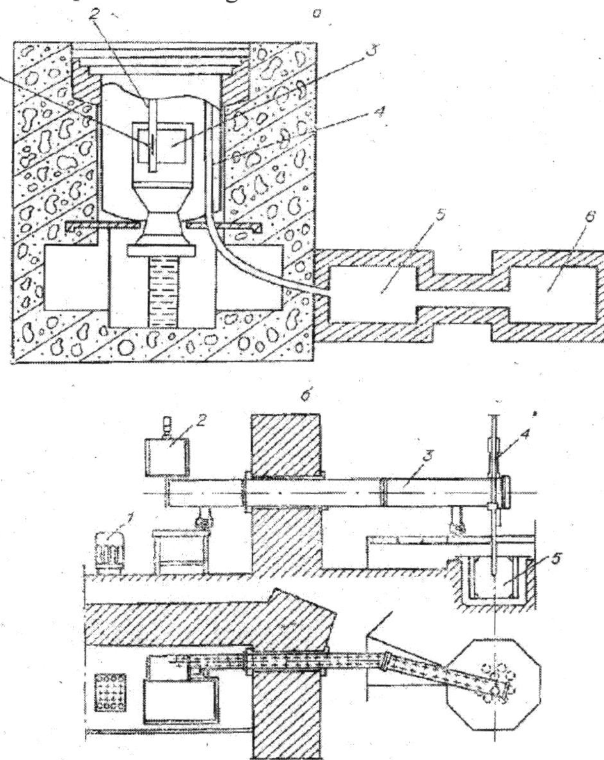

Figure 4.10. Design of a through technological channel with a screen (WWR-t reactor).
a) 1 - a container with a sample, 2 - a working channel, 3 - an active zone, 4 - a transport channel, 5 - a re-chamber, 6 - a decontamination chamber;
b) 1 - transport trolley, 2 - glove box, 3 - conveyor, 4 - lock, 5 - reactor core [4].

Overloading of ampoules from the conveyor into the channel for irradiation and vice versa is done through the gateway. During irradiation, the ampoules are in a stream of heavy water, and after irradiation they are dried in hot helium and fall onto the same conveyor carriage, after which it automatically moves one step. Considering that each ampoule is irradiated for several hours, the total residence time of irradiated ingots on the conveyor is about 2 days, which is sufficient for the complete decay of short-lived radioactive isotopes. A feature of this scheme is the discrete-continuous nature of the irradiation process, in which all operations are automated, except for loading and unloading ampoules with crystals through a glove box, which prevents tritium from entering heavy water through the transport channel into the working room.

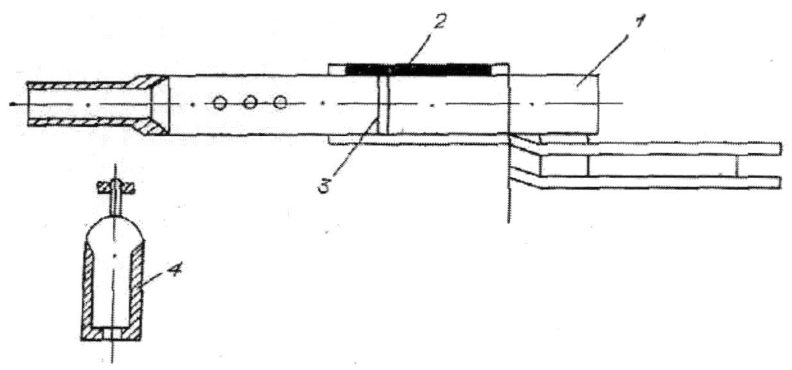

Figure 4.11. Working channel and container for irradiation of ingots with rotation. 1 – duct; 2 – detector; 3 – movable stand; 4 – container.

The dosage of the introduced phosphorus concentration N_p with the static irradiation method can be controlled by the irradiation time, as follows from expressions (2.24) and (2.38), if the neutron flux density φ is assumed to be constant over time.

It does not exclude the possibility of dosing the introduced dopant and by controlling the reactor power at a given exposure time. However, this method eliminates the possibility of conducting independent irradiations in more than one place inside the reactor. On the other hand, the value of N_p, determined by a given concentration of current carriers after doping, also depends on the initial concentration of residual impurities:

$$N_p = \begin{cases} n_i + N_a & (\text{original Si p-type}) \\ n_i - N_d & (\text{original Si n-type}) \end{cases} \tag{4.1}$$

where n_i - concentration of current carriers, providing the desired nominal electrical resistivity of the sample after doping; N_a и N_d - accordingly, the average concentration of residual acceptor or donor impurities in the initial material. Therefore, when determining the required exposure time, the concentration of impurities in the starting material should be taken into account.

In each particular case, the neutron flux density φ in the irradiation zone should be chosen so that the irradiation time is not too long for technical and economic reasons (several hours) and not too small in accordance with the condition

$$t_{irr} > t_l + t_{un} \tag{4.2}$$

where t_l и t_{un} - loading and unloading times, respectively. Condition (4.2) is caused by the fact that all irradiation operations are carried out in a working reactor. Therefore, at short exposure times, when it is comparable with $(t_l + t_{un})$, there is a risk of uneven irradiation, which will lead to uneven introduction of phosphorus atoms along the length of the ingot and the uncertainty in the accumulated dose. This factor is superimposed on the doping nonuniformity discussed above, caused by the distribution of φ along the length and radius of the ingots.

In this regard, both from the point of view of alignment of the inhomogeneity φ along the length of the ingots, and from the point of view of the full use of the working volume of the channels, developed and implemented in the 70 seems more promising in comparison with the static irradiation method at the WWR-t reactor, the method of irradiating ingots during continuous movement (drawing) of them through the working zone of the reactor along the channel axis. When using this method, which requires the presence of through channels for irradiation in the reactor, ampoules with ingots pass through the working zone of the reactor in a continuous stream with a given speed of translational motion and with a given speed of rotation (Fig. 4.12). Dosing the concentration of phosphorus formed is carried out by regulating the speed of translational motion, which is determined by the expression

$$\nu = Ah\varphi_{\text{int}} / \left(n_i \pm N\right) \tag{4.3}$$

where h - reactor working zone height; φ_{int} - integral density of the slow neutron flux, defined as the slow neutron flux that receives an infinitely thin sample when it passes through the reactor working area per unit time; n_i - carrier concentration, which determines the nominal resistivity after doping; N - average concentration of impurities in the initial ingot of p-type (+) or n-type (-); A - the same constant as in (2.38).

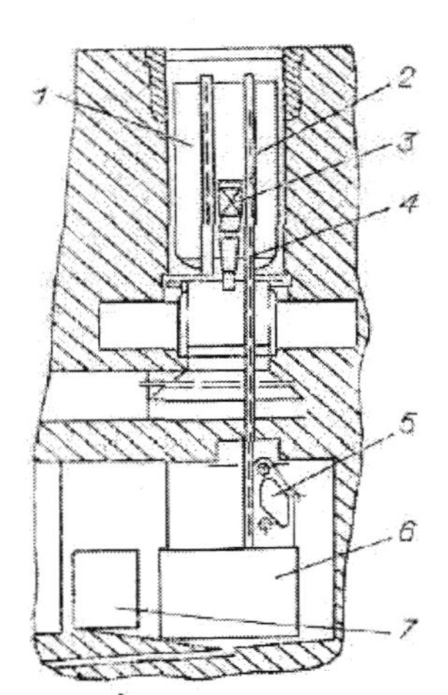

Figure 4.12. Scheme of movement of samples when pulling them through the working zone of the reactor. 1 - preload channel; 2 - working channel; 3 - active zone; 4 - container; 5 - movement mechanism; 6 - receiving hopper; 7- chamber re-adjustment.

After moving through the irradiation zone at a given speed, the ampoules with ingots fall onto the conveyor-settler, which transfers them to the transfer chamber, from where the above described sequence of operations with irradiated material begins. It is quite obvious that using the present method it is possible to effectively ensure a continuous process of doping silicon even in reactors with a large nonuniformity of the neutron flux along the length of the doped ingots (see Fig. 4.4).

All the considered methods of eliminating or taking into account the irregularity of irradiation of doped ingots only allow us to use the calculation method to obtain a given concentration and heterogeneity of the distribution of doping impurities over the volume of extended ingots. To comply with the specified irradiation conditions, for a considerable time it is necessary to equip the reactor with either a standard neutron flux stabilization system that excludes uncontrolled fluctuations in the reactor power and radiation composition at the doping site, or a sufficiently accurate and sensitive system for continuous measurement of the actual flux density and regulation of the actual exposure time.

When organizing the process of nuclear doping, rhodium, vanadium, or other direct-charge detectors that are pre-calibrated according to the doping data of reference silicon samples and located directly at the irradiation sites are currently used as neutron flux sensors. As a result of the development of the corresponding electronic equipment, which integrates the detector current proportional to φ and compares the actual integral neutron flux with the given one over time, it is possible to precisely establish the moment when the specified concentration of dopants is reached. The use of such an independent tracking and control system, which can include an electronic computer, allows you to fully automate the irradiation operation [4] and to achieve a high degree of uniformity and reproducibility of the specified silicon properties inherent in the doping principle [8] over a wide concentration range of the introduced phosphorus.

The algorithm of the optimal arrangement of silicon ingots and the automated calculation of the irradiation regimes during continuous drawing them in the practice of nuclear doping with reference to the WWR-t reactor is implemented as follows.

Following [10], we note that ingots arriving for doping must satisfy certain selection rules. Firstly, in order to achieve maximum performance with a minimum deviation of the final value of the resistivity from the given one, it is necessary that the silicon ingots, at least in the volume of one batch, have irradiation times that fall within the specified limits:

$$T_p - \Delta T_p \leq T_i \leq T_p + \Delta T_p \tag{4.4}$$

where T_p - estimated exposure time, ΔT_p - permissible deviation from the estimated time of irradiation of an individual ingot, T_i - single ingot irradiation time. In turn, T_i can be found from the expression:

$$T_i = 2{,}156 \cdot 10^{19} \left(\rho_i - \rho_k\right) / \varphi_{\text{int}} \rho_i \rho_k \tag{4.5}$$

where ρ_i - average value of resistivity in an n-type conductivity ingot before irradiation, ρ_k - final value of resistivity.

Secondly, alloyed silicon ingots having different lengths should be placed as tightly as possible in containers of standard length. Currently, technical specifications allow the supply of silicon ingots with a length of 60 mm and above. The standard container length is 500 mm. Therefore, irradiated ingots must be placed in containers so that their total length is as close as possible to the standard container length.

In addition, in the practice of doping, it often becomes necessary to group the containers in such a way that the total irradiation time does not exceed, for example, the weekly cycle of the reactor at optimal power.

It is obvious that in the conditions of mass production of NDS, the solution of such a problem as choosing the optimal irradiation mode, arranging ingots in batches and grouping them by containers, as well as forming weekly irradiation cycles, is expediently carried out using a computer. For this, as you know, the creation of its mathematical model is necessary. We formulate the problem. There are cylindrical objects P_i $(i = l, n)$ of height h_i, diameter d_i, mass p_i, and also the parameter ρ_i, which determines the mode of their irradiation T_i.

Necessary:

1) from all the T_i modes $(i = l, n)$ determine the optimal ${}^{opt}T_k$ mode $(k=1, 2, 3, ...)$ that corresponds to the group of objects P_k $(1 \le k \le n)$ selected in a certain way;

2) place P_i objects in the cylindrical containers Ω_r $(r=1, 2, 3, ...)$ so that the unoccupied part of the areas Ω_r is the smallest. H_r and D_r are the sizes of the region Ω_r, r is the number of containers for placing all objects P_i;

3) for n_p containers with ${}^{opt}T_k$ mode, group Ω_r containers so that their number is equal to or a multiple of the specified number n_k for a weekly work cycle. ${}^{opt}T_k$ modes $(k=1, 2, 3, ...)$ are found as follows. For each object P_i, T_i is calculated by the formula (4.5). The found T_i values are arranged in increasing order $(T_i \le T_{i+1})$. Arrays of parameters h_i and p_i are also constructed. Each T_i value is associated with elements of the array N_i $(i= 1, n)$ from the natural series of numbers from 1 to n. An example of array construction is given in table 4.1.

Matrix Φ_{ij} $(i,j=1, n)$ built on the following principle:

$$\Phi_{ij} = \begin{cases} 1, \text{ if } T_i - \Delta T_j \le T_i \le T_j + \Delta T_j \\ 0, \text{ otherwise} \end{cases}$$

Obviously, the elements φ_j of the matrix Φ_{ij} are the numbers N_i (i=l, w) of the objects P_i (i=1, w), and each j-th column is a set of numbers of those objects P_i (i=1, w) for which T_i corresponds to condition (4.4), w - is the number of nonzero elements of the column φ_j.

In the matrix Φ_{ij} defined column φ_k, for which $a=\max\{a_j\}$, where $1 \le j \le n$; a - the total length of all ingots described by nonzero numbers of the j-th column, which will ensure the most complete load of equipment. The mode corresponding to this column is selected as optimal $({}^{opt}T_k)$. If the value of a is not the only one, then in order to save the energy resources of the reactor, the mode $^{опт}T_k = \min\{^{opt}T_j\}$.

Table 4.1. Array of object parameters

№	h_i, mm	P_i, g	T-0,05T, min	T, min	T+0,05T, min
1	90	470	1011	1064	1117
2	158	848	1030	1084	1138
3	252	1300	1038	1092	1147
4	115	615	1038	1093	1148
5	86	450	1039	1094	1148
6	151	817	1046	1101	1156
7	74	378	1048	1104	1159
8	255	1346	1059	1115	1170
9	233	1256	1060	1115	1171
10	266	1392	1061	1116	1172
11	153	816	1061	1117	1173
12	215	1122	1063	1119	1175
13	237	1298	1063	1119	1175
14	112	604	1076	1136	1192
15	270	1442	1080	1137	1194
16	281	1486	1089	1147	1204
17	260	1398	1093	1150	1208
18	273	1457	1098	1156	1214

The problem of arranging the objects P_i ($i=l, n$) in the containers Ω_r (r= 1, 2, 3...) can be considered as the problem of optimal placement of geometric objects. We give its mathematical formulation. For this, we choose the poles of the cylindrical objects P_i and the object Ω_r, the centers of their lower bases [11]. We place Ω_r so that its pole coincides with the beginning of the Cartesian coordinate system *XYZ*, and its generator is parallel to the *Z* axis.

In the problem posed, the objects P_i and the container Ω_r have close diameters, and their generators are mutually parallel. Then, according to [11], the mathematical formulation of the problem in this version will include:

a) terms of placement of objects P_i in a given area Ω_r:

$$0 \le Z_i \le H_r - h_i \ (i = 1, n) \qquad (4.6)$$

where Z_i the essence of the cartesian coordinate of the pole of the object P_i,

b) conditions of mutual non-intersection of objects P_i and P_j;

$$Z_i - Z_j = \begin{cases} h_i, \text{if } Z_i - Z_j < 0 \\ h_j, \text{if } Z_i - Z_j > 0 \end{cases} \quad (i, j = 1, n; i \neq j) \qquad (4.7)$$

c) goal function [11], which has the form:

$$K_r(Z) = \underset{i=t}{\overset{1}{\vee}} (Z_i + h_i) \quad (1 \le t \le w)$$
$$(1 \le w \le n; 1 \le l \le w; r = 1, 2, 3...) \qquad (4.8)$$

where V - R-disjunction operation [11].

Thus, the formulated problem reduces to finding the values of $K(Z)$: $Z \in G_o$, where G_o is determined by the system of inequalities (4.6) and (4.7). Taking into account the features of such problems noted in [12], as well as the conditions for the formulation of the real problem, we can obviously confine ourselves to finding rational values of the objective function $K_r(Z)$ that are close to optimal within the given accuracy of approximation to the boundary of the unoccupied part of the domain Ω_r (r=l, 2, 3...). It should be noted that when searching for the next value of the target function $K_{r+1}(Z)$, those objects that were included in the search for the previous layout $K_r(Z)$.

The determination of local values of function (4.8), i.e., the selection of rational layouts of objects in containers, is implemented in the form of an algorithm, a block diagram of which is shown in Figure 4.13.

1. We determine g= H_r-K_r(Z) as an unfilled part of the area Ω_r (r=l, 2, 3...).
2. Of all the elements N_k(I≤k≤w) the column φ_k all kinds of combinations are determined ^mC_w (m=1, 2, 3...l; 1≤ w ≤ n). The value of l is set from the condition of impossibility of assembling objects P_i(i=l, I),

   $$K_m = \underset{i=1}{\overset{l}{\vee}} (Z_i + h_i) \gg H_r$$

   To store and process all values ^mC_w [no less (w/m)=w!/(w-m)!m!] requires a large amount of RAM and a significant amount of computer time. They can be significantly reduced if, out of the total possible combination of combinations of objects, 2, 3 ... m (m=l, 2, 3...l) for each object P_i (i=l, w) we remember only the best fillings $_1K^t$ (t=l, w).
3. From an array of values $_1K^t$ layouts are selected $_2K^S$ (s≤t) which contain objects with a total height (H_{r-g}), moreover, preference is

given to those layouts that include objects with modes closest to ${}^{opt}T_k$,

4. Matching set ${}_2K^S$ number combination N_s fixed in the column of results Ψ_0.
5. From the matrix Φ_{ij} the elements of the result column are deleted in Ψ_0.
6. From the array ${}_1K^t$ those elements that correspond to the combination of numbers, including at least one of the elements of the column, are deleted Ψ_0. Resulting matrix Φ_{ij} (and column j_k) will be filled only with those object numbers P_i, which are not in the layouts ${}_2K^S$.
7. The calculation procedure (go to step 3) is repeated until the array ${}_1K^t$ at least one nonzero element remains. Otherwise, it is necessary to proceed to step 2. And if, after its execution, the array ${}_1K^t$ will remain empty, go to the next paragraph of the algorithm.
8. Change the border of the unoccupied part of the region Ω_r with a given step e: $g_b=(b-1)e$, where b - iteration number.
9. We check the fulfillment of the conditions:

$$\varphi_{ik} = 0\left(i = 1, n\right); g_b > g_{\max} \tag{4.9}$$

where g_{Max} - boundary of the empty space of the region admissible by the condition of the problem Ω_r.

When one of the conditions (4.9) is fulfilled, the layout process is completed (paragraph 10). Otherwise, we proceed to calculations, starting with step 2.

10. We carry out the grouping of containers for a weekly cycle of work. For mode ${}^{опт}T_k$ determine the required number of containers n_k (k=1, 2, 3, ...) according to the formula:

$$T_c \geq \left(n_k + 2\right)T^l \tag{4.10}$$

where T_c - reactor cycle time, T^l - travel distance equal to the standard length of the block container.

We determine the remainder of containers n_r, not included in the integer for weekly loading cycles, i.e.

$$n_r = \left(n_p - n_k\right)E\left(n_p / n_k\right) \tag{4.11}$$

where $E(n_p/n_k)$ - whole part of the relationship, n_p - total number of mode containers ${}^{opt}T_k$.

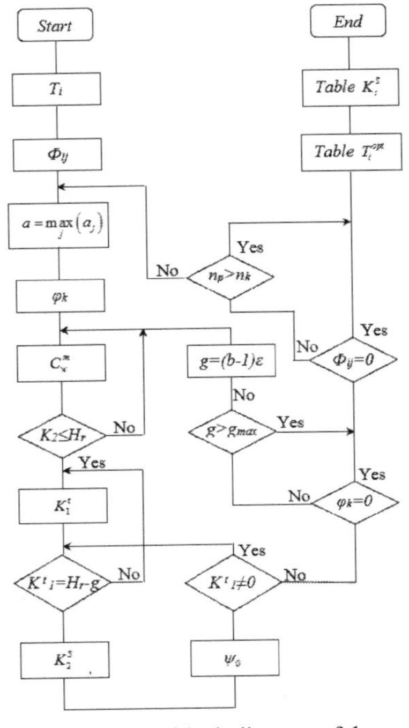

Figure 4.13 block diagram of the algorithm for solving problems

In the case when $n_p \geq n_k$, leave n_p-n_r the best (in the sense of filling the area Ω_r and proximity to the regime ${}^{opt}T_k$) ingot containers. The objects P_r (r=1, 2, 3, ...), arranged in the remaining containers, leave in the matrix Φ_{ij} for further processing and go to the definition of the mode ${}^{opt}T_{k+1}$. Otherwise, the result is all the resulting layouts. The process of solving the problem ends if the matrix is trivial, or the remaining objects are not compiled within the permissible value of the unfilled part.

Table 4.1 shows an array of parameters of 18 objects, and table 4.2 shows the found layouts of these objects for optimal mode (${}^{opt}T$=1098 min).

In conclusion, we note that the implementation of the algorithm allows not only to eliminate the routine calculation of the irradiation modes, but also significantly increase the productivity of the irradiation unit as a result of a more rational layout of the ingots and an improvement in the grouping of containers in the case of a weekly cycle of operation of a nuclear reactor at optimal power.

Table 4.2. The result of the layout of objects (Hr = 500)

Ingot combinations				H_r-g, mm	ΣP, r
10	-	-	9	499	2648
16	-	-	12	496	2608
17	-	-	13	497	2696
2	3	-	1	500	2618
14	18	-	4	500	2676
7	15	-	6	495	2637
8	11	-	5	495	2612

4.3 Specialized Nuclear Reactor

Known data on the irradiation conditions, properties and scale of production of nuclear doped silicon at various reactors of the world are given in table 4.3.

In world practice, until recently, the production of NDS is tied to existing reactors. However, the current tendency toward an increase in the output of NDS makes it economically feasible to create special devices for irradiation, for example, in the form of a specialized atomic reactor (SAR). This task becomes especially urgent in connection with the tightening of requirements for the properties of NDS, as well as in connection with an increase in the diameter and length of alloyed ingots. Dimensions of ingots with a diameter of up to 30 cm and a length of up to 150 cm must be considered as already established practice. When implementing the process of nuclear doping, it is also necessary to ensure optimal exposure conditions.

Below, by calculation, the possibility of solving the problem of nuclear doping of silicon in atomic reactors [6] of an optimal design is evaluated. In the calculations, the irradiation of silicon ingots in a nuclear reactor is represented by two idealized schemes: in the first scheme, a silicon cylindrical ingot with a diameter of 15 cm is located in the center of the heat column surrounded by an annular active zone (Fig. 4.14), in the second, the silicon layer is located in the reflector at a certain distance from a cylindrical core with a diameter of 30 cm (Fig. 4.15). In both irradiation schemes, the core consists of uranium-235, aluminum and light water. The active zones of this composition are notable for their small geometric dimensions and relatively low cost. In the calculations, graphite, beryllium, and heavy and light water were considered as moderating materials of the heat column and reflector.

The prospects of using moderators for nuclear doping of silicon in reactor conditions were evaluated by such indicators as the efficiency of using slow neutrons, the content of fast neutrons in the total neutron flux, and irregularity of the ingot irradiation. In accordance with this, the following were calculated: K_{isp} - neutron utilization coefficient equal to the ratio of the number of acts of capture of slow neutrons by a silicon isotope ^{30}Si to the number of nuclear fission acts in the core, φ_f/φ_s - ratio of fast and slow neutron fluxes, $(\varphi_{max} - \varphi_{min})/(\varphi_{max} + \varphi_{min})$ uneven distribution of the flux density of slow neutrons over a silicon ingot. In multi-group calculations

Table 4.3. The conditions for the production of nuclear doped silicon (NDS) in various reactors

Reactor type	Moderator	Neutron flux density, $cm^{-2} \cdot c^{-1}$	The ratio of thermal and fast neutron flux	Maximum dimensions of irradiated ingots, mm	Irregularity of exposure along the length δl and radius δr	The nature or volume of production of NDS	References
Reactor ESSOR Joint Research Center in Ispra (Italy)	D_2O	$(2,7\text{-}3) \cdot 10^{14}$	>400	Ø up to 77 l up to 500	$\delta l = \pm 4\%$	Mass production	[4]
University of Missouri MURR Research Reactor	H_2O	$6 \cdot 10^{14}$, c. $10^{13}\text{-}10^{14}$, g.f.	8:1, c. 20:1-10:1, g.f.	Ø up to 86 l up to 250	$\delta l = \pm 4\%$ $\delta r = \pm 1\%$	15 t/year based on ρ=50 Ohm·cm	[5]
DIDO and PLUTO Research Reactor in Harwell (England)	D_2O	$\sim 10^{13}$, des. $\sim 10^{14}$, impl.	1000:1 des. 100:1, impl.	Ø up to 100 l up to 550 Ø up to 115 l up to 600	$\delta l < 5\%$ $\delta r < 4\%$	15 t/year 20 t/year	[9]
Reactor National Bureau of Standards (USA)	D_2O	Up to 10^{14}	Cu-Cd ratio 46-3400, c.	Ø up to 100		Experienced parties	[10]
General Electric GETR Test Reactor (USA)	H_2O	Up to $1{,}5 \cdot 10^{14}$	~10:1	Ø up to 82	$\delta l = \pm 5\%$ $\delta r = \pm 1\text{-}3\%$	Planned hundreds of kilograms per month	[11]
Research reactor VVr-ts (USSR)	H_2O	$\sim 10^{14}$, c.	~10:1	Ø up to 86 l up to 200 (static mode), l up to 500 mm (drive mode)	$\delta l < 5\text{-}7\%$ $\delta r < 4\%$	Mass production	Authors data

Note: c. – core; g.f. – graphite reflector; des. – design parameters; impl. – implemented parameters performed on a computer, the library of constants BNAB-78 was used [16].

In Fig. 4.14 presents the dependence of the value K_{isp} from the width of the layer of the heat column D (cm) in the first irradiation scheme, in Fig. 4.15 - dependence of the value K_{isp} from the distance of the annular silicon layer to the active zone in the second irradiation scheme. In the table. 4.4 shows the values corresponding to the maximum values K_{isp}.

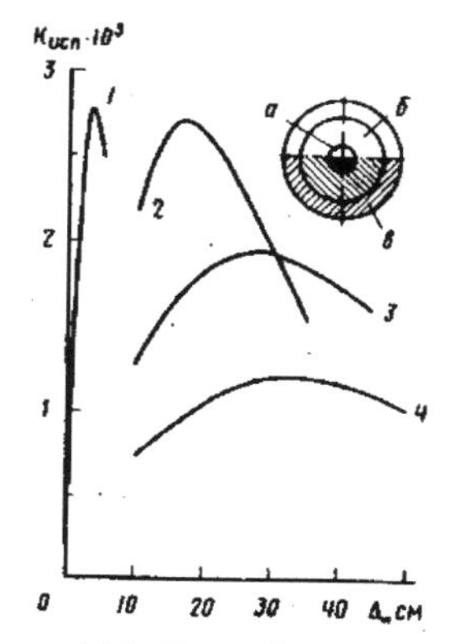

Figure 4.14. Dependence of the K_{isp} value on the moderator layer width in the first irradiation scheme:
1 - light water; 2 - beryllium; 3 - heavy water; 4 - graphite; a is an ingot of silicon; b - moderator; c - core

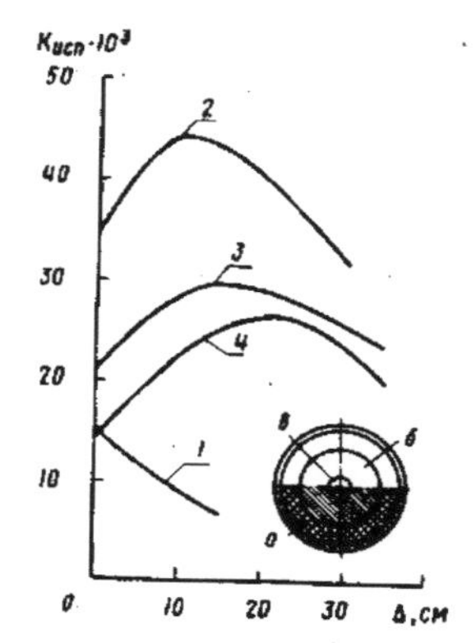

Figure 4.15. The dependence of the value of K_{isp} on the distance of the annular silicon layer to the active zone in the second irradiation scheme:
1 - light water; 2 - beryllium; 3 - heavy water; 4 - graphite; a - annular layer of silicon; b - reflector; c - core

Obviously, in the first scheme, due to the symmetry of irradiation, the uneven distribution of the flux density of slow neutrons across the ingot will be minimal, since the non-uniformity in this case is determined only by the distance and absorption of neutrons in the ingot. The value of K_{isp}, which determines the efficiency of using fuel in a reactor during nuclear doping, is small in this irradiation scheme, about 0.3%. In the second irradiation scheme, the K_{isp} value is approximately an order of magnitude higher than in the first scheme. A similar relationship is observed for the uneven distribution of the density of the flux of slow neutrons in the bulk of the ingot, which in this case depends mainly on the gradient of the density of the neutron flux along the radius of the reflector at the location of the ingot.

From table 4.4 it follows that beryllium, which is significantly superior to other moderators, is inferior to the non-uniformity of irradiation in the second scheme in graphite and heavy water [17] as a retardant material for a heat column and a reflector.

The positive qualities of both irradiation schemes can be combined in the version of the reactor, the active zone of which consists of three modules. In the space between the modules, as shown by the calculations, the skew of the slow neutron flux density will be two to four times less than in the second irradiation scheme. The SAR power for the production of NDS can be small, about 30 MW and, accordingly, the specific core energy density is also low (0.1 MW/l), which allows the use of already developed and operated fuel elements.

It should be noted that for the construction of SAR on an individual site, an economic justification is required taking into account the prospects for 20-30 years. In the case of small or difficult to predict needs, such a justification will not be convincing enough. Therefore, naturally, the desire to use existing or under construction reactors for doping purposes, despite the fact that the quality of the products produced due to the already established reactor characteristics, will be lower than in the case of production on superlattices.

When using nuclear power plants (NPPs) for doping silicon, two independent tasks must be solved.

One of them is related to the fact that there are no ready-made technological channels in the nuclear power plant, they need to be designed and, possibly, made adjustments to the design of the reactor itself, which can only be done in a newly constructed or planned installation. The dimensions of the technological channels and their position in the reactor cannot be selected taking into account all the necessary requirements, since this is to some extent limited by the existing reactor design.

Another problem arises due to the fact that the irradiation mode of the samples is forced to adapt to the variable operating mode of the nuclear power plant. In this case, it is difficult to control the dose of radiation and the choice of the mode of operation of the technological channel with continuous loading of ingots for irradiation. In a high power reactor with an active zone geometry (N / D) <1, i.e., when the height of the active zone is less than its diameter, the error in monitoring the radiation dose can reach

10-15% due to the neutron field mobility inherent in these reactors if standard reactor equipment is used for this. An example of this is the RBMK type reactor.

As calculations showed, the best reactor design for irradiating silicon is one in which the core geometry corresponds to (H / D)> 1. In this design, in addition to eliminating the indicated uncertainty in the level of the slow neutron flux, the irregularity of the distribution of the neutron flux along the radius of the reflector is significantly reduced. The listed limitations and disadvantages when using nuclear power plants for the production of NDS are absent in SAR, for example, in the variant proposed above. The implementation of SAR can be economically justified if such a reactor is built at high power NPPs using the main and auxiliary nuclear power systems and for SAR.

The technical feasibility of such a project is determined by the insignificant ratio of the power SAR and modern NPP. When constructing a SAR in the form of a satellite reactor of a nuclear power plant, savings of about 80–90% of the costs of constructing a SAR in an autonomous design can be achieved. The cost of the production line in the production of NDS is almost the same in any variant of the organization of silicon irradiation.

Table 4.4. Indicators of retarding materials for a heat column or reflector

Moderator	Maximum value, K_{isp} x10^3		Layer of heat column or reflector		Correlation of fluxes of fast and slow neutrons		Irregularity of irradiation of silicon ingots, rel. units	
Silicon ingot irradiation schemes								
	1	2	1	2	1	2	1	2
Graphite	1,2	27	33	23	0,88	0,96	0,03-0,04	0,18
Heavy water	0,	30	28	16	0,39	0,59	0,03-0,04	0,21
Beryllium	2,8	45	17	11	0,75	0,43	0,03-0,04	0,25
Light water	2,8	15	3,5	0	1,12	0,58	0,03-0,04	0,38

In the proposed SAR project in the form of a high-power nuclear power reactor satellite, there are significant advantages over the possible options for using existing reactors, both in terms of the quality of nuclear fuel and the speed of its production (table 4.5). Despite the fact that the cooling system of the nuclear power plant is used, the autonomy of the SAR is not violated, and therefore, precise control of the radiation dose can be carried out.

Thus, it seems to be the most rational to organize the production of nuclear-fuel complex on a specialized high-power nuclear power reactor-satellite.

In conclusion of the section, we note that at present, the volume of production of NDS in the world reaches about 150 tons / year. In the Soviet Union, the production of NDS was carried out depending on the purpose of TU 48-4-430-81 and TU 48-0513-12 / 0-81 for consumers in the electronic and electrical industries.

Table 4.5. Indicative indicators of the use of atomic reactors for nuclear weapons

Type of nuclear reactor	The content of the organization of production	The relative cost of work	Performance	Possibility of expanding the range of doped ingots		Production organization term
				at face values ρ	in diameter and length	
Nuclear power plant	Design change	0,1-0,3	Limited	Limited	Limited	3-7 years
SR	Design and construction of the complex	1	Not limited	Not limited	Not limited	5-7 years
NPP + SR	Design and construction of a reactor	0,1-0,3	The same	The same	The same	3-5 years

Note, ρ-final resistivity of the NDS.

4.4. Annealing

The hardware design of the annealing operation of nuclear doped silicon is standard. A conventional industrial diffusion furnace of the same type as in the manufacture of semiconductor devices is used. It has a fairly long section with minimal temperature fluctuations. In addition, the furnace is equipped with a temperature stabilization system and a system for creating the necessary annealing medium that is quite suitable for annealing.

As for the annealing process itself, there is currently no single approach to its implementation in the literature. The annealing modes, which are given in a number of original works, differ significantly. In particular, the annealing temperature ranges from 500 [18, 19] to 1100°C [20], the holding time is from minutes [4, 21] to several tens of hours [19]. In addition, annealing is carried out in a wide variety of atmospheres (air, vacuum, neutral gases, neutral gases with the addition of HCl etc.), in containers of different materials (quartz, silicon, etc.)

To date, annealing is the least scientifically substantiated stage of obtaining NDS, and technologists involved in NDS are guided when carrying out annealing more by their experience and intuition than by strict scientific criteria relating to some abstract silicon.

One of the reasons for the significant differences in the conditions of annealing of the NDS is the different criteria for assessing the completeness of annealing of radiation defects (see Chapter 3). Another reason is due to the difference in irradiation conditions and the history of the samples. Indeed, as shown in Chapter 3, the observed changes in the electrophysical, optical, and mechanical properties of single-crystal silicon are associated mainly with the appearance of secondary radiation defects - divacancies, A-centers, E-centers, disordered regions, etc. during irradiation and transformation during annealing. In addition, both during irradiation and during annealing, primary and secondary defects actively interact with impurities, thermal defects, dislocations, interfaces, etc. The observed effects also depend on the irradiation conditions — temperature, radiation intensity, ratio of radiation components, and others. In this case, concomitant radiation defects are formed in crystals, which are characterized by a certain set of residual electrically active and inactive impurities, as well as thermal defects and structural defects formed during crystal growth. A similar situation occurs with nuclear doping, but with a number of differences. So, monocrystalline silicon is subjected to irradiation with a minimum content of residual electrically active impurities (boron, phosphorus). Radiation defects accumulate simultaneously with the formation of dopant atoms of phosphorus.

During doping, silicon crystals are irradiated with sufficiently large doses of reactor neutrons. This affects the formation of defects and their annealing, on which the electrophysical properties of nuclear doped silicon ultimately depend. Despite the well-known ambiguity of the data on the optimal modes of annealing of NDS, it can be fairly confidently stated that as a result of annealing at temperatures of 700-800° C for 0.5-2 hours, the observed radiation disturbances are almost completely eliminated with obtaining stable values of resistivity, concentration and mobility of charge carriers [22–25], however, for reasons that are not yet clear enough, in some cases, to obtain stable values of the lifetime of minority carriers, annealing at higher t temperatures [17, 18, 26].

The main laws of isochronous annealing, controlled by the change in the resistivity of silicon samples obtained by the crucibleless zone melting method and irradiated by reactor neutrons at a moderator temperature (50°C), are shown in Fig. 4.16 [22]. After irradiation, the samples are characterized by electrical resistivity ρ close to its eigenvalue (see also [23]). As the annealing temperature increases, the curves exhibit features related to changes in the structure and charge state of radiation defects. Curve *1* reaches saturation after annealing at a temperature of about 700°C, the samples have an electronic type of conductivity, and the carrier concentration in them is close to the calculated values.

Now let us consider how the annealing parameters can change when the irradiation conditions and the background of the initial single-crystal silicon change. As an example of the influence of the "rigidity" of the neutron spectrum on the annealing parameters, curve *2* in Fig. 4.16. The data were obtained for similar samples irradiated so that the rate of accumulation of radiation defects remained unchanged, and the rate of phosphorus formation decreased by about 2 times. It is seen that the nature of the change in resistivity depending on the annealing temperature remains the same.

However, in the samples irradiated with a more "hard" spectrum, a shift of the conversion point towards higher temperatures is observed, respectively, the saturation exit region ρ shifts in the same direction. The authors of [26] come to the same conclusion based on the data presented in Fig. 4.17 and those obtained in the study of the recovery of ρ and the lifetime of charge carriers in silicon samples irradiated by a flux with a different ratio of thermal and fast neutrons. It turned out that when the indicated ratio changes by more than two orders of magnitude, the annealing temperature shifts by 70–100°C.

Noteworthy is the rather high annealing temperature, which reaches 800°C and higher, even in the case of irradiation with practically only thermal neutrons [4, 27]. These facts can be explained as follows. Recall that when irradiated with reactor neutrons, a whole spectrum of various disturbances is formed, the most complex of which are defects that arise when slowing down initially displaced silicon atoms that receive energy from fast neutrons. Most likely, these defects are disordered regions. According to [28], annealing of individual disordered regions introduced by ion irradiation occurs at temperatures of about 300°C. Apparently, the same is true for neutron irradiation.

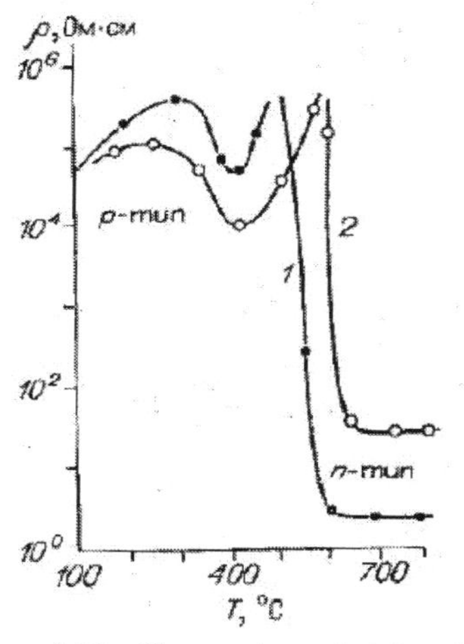

Figure 4.16. Change in resistivity during isochronous annealing of silicon irradiated with a full spectrum of reactor neutrons (1) and a spectrum with a cutoff of slow neutrons (Cd screen) (2).

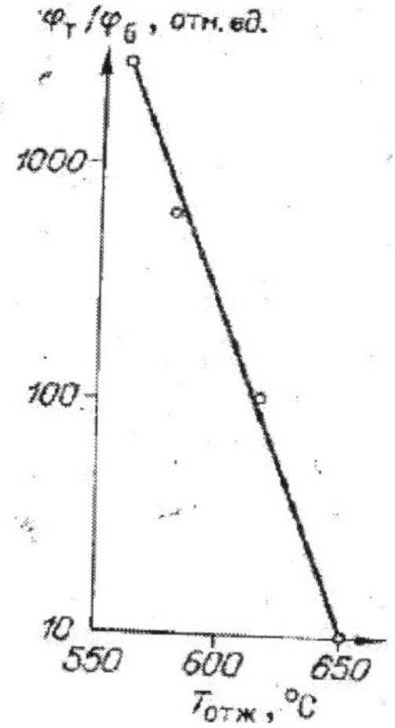

Figure 4.17. Change in the temperature of silicon annealing depending on the rigidity of the spectrum of reactor neutrons [26].

Indeed, in the IR absorption spectra characteristic of silicon samples irradiated with reactor neutrons at temperatures of 50–70° C, significant changes are observed in the first stages of annealing, which can be associated with the decay of disordered regions. Near the edge absorption disappears, the intensity of the absorption bands in the region of 18-24 μm decreases noticeably (Fig. 4.18) [29]. This process is accompanied by the formation of defects already introduced during the irradiation process, and the formation of new ones, which, as the annealing temperature rises to 450–500° C, begin to dominate. The latter is clearly seen on the curves of isochronous annealing (see fig. 4.16).

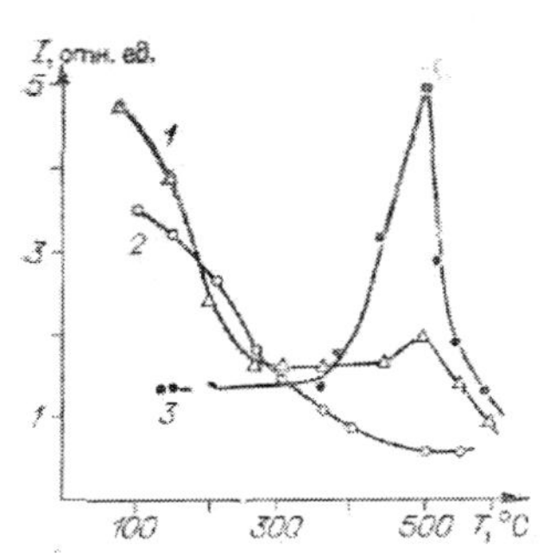

Figure 4.18 Change in the intensity of IR absorption bands during isochronous annealing. 1-2; 2 - 20.5; 3 - 9 microns

After annealing at temperatures of 400–500° C, the maximum conductivity is achieved due to defects with a low level of occurrence exhibiting acceptor properties [30]. Moreover, in the IR spectra in the range of 4.5-14.5 μm, a complex absorption appears, including about 20 distinct peaks (Fig. 4.19).

Absorption disappears after annealing at temperatures of 600-650° C. In the same range on the curves of fig. 4.16 conductivity type conversion is observed. Thus, disordered regions to a certain concentration do not determine the annealing process. They serve as the pantry, which stores a huge number of point defects. The fine structure of such formations was considered in [23]. It has been shown that individual elements of disordered regions include clusters of point defects formed as a result of cascade displacement of atoms from lattice sites in the region of tracks of initially knocked out silicon atoms.

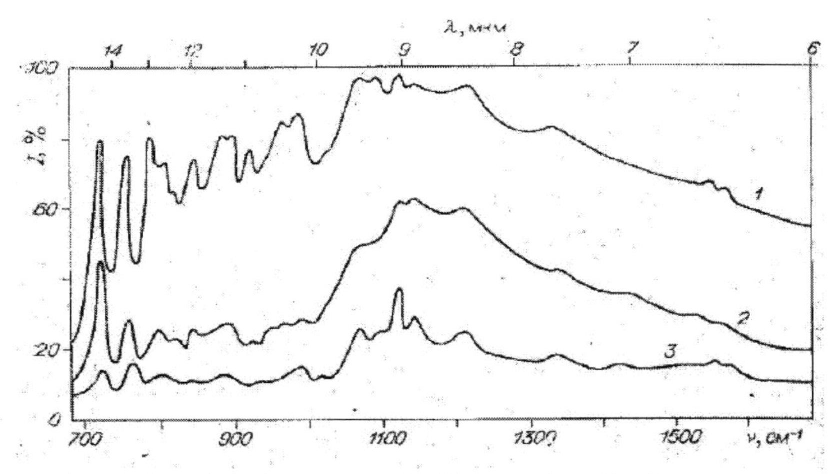

Figure 4.19. IR absorption spectrum of a silicon sample irradiated by neutrons at different stages of isochronous annealing. 1-500; 2-550; 3- 650°C.

It is natural to assume that disordered regions play the role of nucleation centers for a new type of defect. This assumption is supported by the data of [31], which showed that hole conduction is clearly observed after irradiation with fast electrons in the same range of annealing temperatures. Note that upon irradiation with fast electrons with an energy of 10 MeV, disordered regions do not form in silicon. A return to the initial values of specific resistance in these samples is observed after annealing at temperatures of about 700° C. Consequently, changes in properties similar to those occurring during annealing of silicon irradiated with neutrons can also be achieved by introducing only point defects into silicon.

Residual impurities in the initial silicon have a more significant effect on the annealing parameters. Under similar irradiation and annealing of silicon with a high oxygen content, the removal of defects is delayed to 1000–1200° C [31–33]. On the Figure 4.20 shows the recovery curves for the intensity of neutron diffraction reflections (004) and (220) during isochronous annealing of silicon grown by the Czochralski method and irradiated with reactor neutrons at a temperature of 320 K [31]. It is noteworthy that the restoration of intensity begins only after annealing at a temperature of about 600° C and ends at temperatures of about 1200° C.

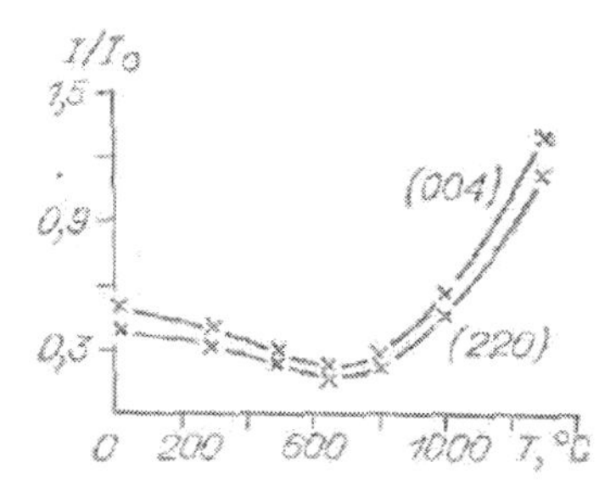

Figure 4.20. Change in the intensity of peaks of elastic coherent neutron scattering during isochronous annealing of a silicon single crystal irradiated by reactor neutrons.

In this connection, data [24] are of interest that dislocation loops (~ 10^{14} cm^{-3}) with a diameter of 50–100 angstroms, which possibly participate in carrier capture, are retained in silicon irradiated with neutrons up to 900° C, decreasing time of their life. It is also possible that they are detected during neutron diffraction studies.

It should be noted that the neutron diffraction method used in [31] is very sensitive to the presence of defects in single-crystal silicon. Since the peak intensity of reflections is determined by the square of the concentration of scattering centers, any factors that can change the size and concentration of these centers will strongly affect the peak intensity of reflections [34]. From this point of view, the deformed sections of the crystal lattice in the region of defects can be considered as giving an additional contribution to incoherent scattering. In addition, this method is convenient for studying structural defects, including radiation ones, because unlike EPR, IR spectroscopy, and others, it is insensitive to the charge state of defects, which, together with other methods, provides valuable information on the transformation mechanism defects during irradiation and during heat treatment.

The role of structural defects as sinks for radiation defects and their annihilation centers has already been noted. Obviously, structural defects can noticeably affect the annealing parameters.

In [39], *n*-type silicon samples without dislocations and with a dislocation density of $5 \cdot 10^4$ cm^{-2} were used for studies. Single crystals were grown at different rates by the crucibleless melting method in vacuum and in an Ar + H_2 mixture, which determined the type and concentration of microdefects in dislocationless silicon [36]. The characteristics of some of the investigated crystals are given in table 4.6. Samples in the form of plates with a thickness of 4 mm were irradiated with various doses of reactor neutrons in cadmium cases and without an absorber. Annealing was carried out in vacuum at various temperatures up to 1100°C. In fig. 4.21, and the theoretically calculated increment of conductivity $\Delta\sigma$ is compared with the measured values in silicon samples irradiated with different doses and undergoing subsequent high-temperature annealing. The concentration of introduced phosphorus was calculated according to (2.24). The experimental values of $\Delta\sigma$ depending on the dose vary linearly, at least in the studied range, but their absolute values for all crystals are less than calculated. For different samples, the angle of inclination of the curves is also different. We note that all samples were irradiated and annealed under identical conditions, and the experimental values of $\Delta\sigma$ were given after annealing at such temperatures when its further increase did not change the conductivity.

The largest deviation $\Delta\sigma$ from the calculated one is observed for dislocation crystals grown in vacuum at a rate of 8 mm / min (curve *1* in Fig. 4.21, a). In similar samples, but grown at a speed of 2-3 mm / min, the deviations $\Delta\sigma$ are less (curve *2* in Fig. 4.21 a). This difference can be explained by the fact that, at low growth rates, impurities have time to condense in the region of dislocations. In crystals grown at a rate of 8 mm /

min, diffusion of impurities to growth defects is limited. If we take into account that these crystals contain little impurity, we can assume that the dislocations in them are more "pure" and some of the phosphorus atoms formed as a result of nuclear transmutations will be deposited on them.

Table 4.6. Characterization of the structure of single-crystal silicon grown by the crucibleless zone melting method under different conditions.

Sample No.	Speed, mm/min and growing medium	The density of dislocations, cm^{-2}	Resistivity, Ohm · cm
1	8 (vacuum)	10^4	$6 \cdot 10^3$
2	2,2 (vacuum)	$5 \cdot 10^4$	180
3	3 (90%Ar+10% H_2)	0	260
4	2 (90% Ar +10% H_2)	0	90
5	2 (90% Ar +10% H_2)	0	190
6	3,2 (vacuum)	0	50

Minimum deviations $\Delta\sigma$ are characteristic of crystals grown in an Ar + H_2 atmosphere and containing an insignificant concentration of growth microdefects (curves *3*, *4* in Fig. 4.21, a). The opposite effect is observed in the initial sections of the curves for some crystals - the experimental values exceed the calculated ones. It can be assumed that during irradiation and subsequent annealing, the electrical activity of some of the phosphorus atoms that are electrically inactive in the initial crystals is restored.

This assumption, in our opinion, is well confirmed by the experimental results (Fig. 4.21, b). In this case, to reduce the rate of introduction of phosphorus, the samples were irradiated in cadmium canisters. The course of the change from the dose differs from the linear one (except for curve *1*). At the beginning, the quantity $\Delta\sigma$ increases sharply, then passes through a maximum. This effect, which is especially noticeable in dislocation silicon grown at a speed of 2-3 mm/min, is practically absent in dislocation silicon grown at a speed of 8 mm/min. The dose dependence of $\Delta\sigma$ for this crystal is practically linear.

Apparently, at low growth rates, as already mentioned, part of the phosphorus forms electrically inactive complexes in the region of growth defects, which are destroyed by irradiation and subsequent annealing, which leads to an increase in the conductivity of crystals. Since sample 1 (see table 4.6) is characterized by a low dopant content in general, this process is absent in it.

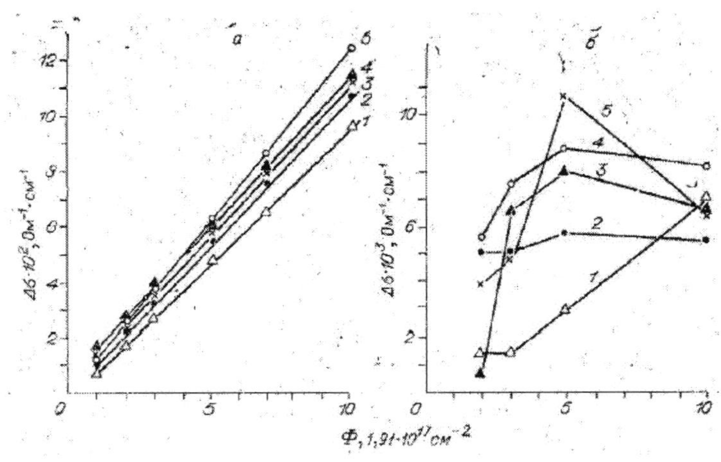

Figure 4.21. Dependence of the change in conductivity on the degree of defectiveness of silicon crystals irradiated with neutrons, without cadmium protection (a) and in the cadmium case (b). The numbers of the curves correspond to the numbers samples in the table. 4.2. in the following way. a) *1* - 1, *2* - 6, *3* - 3, *4* - 4, *5* - theory; b) *1* - 1, *2* - 3, *3* - 4, *5* - 6.

Thus, the deficit or excess of the concentration of electrically active phosphorus compared to the expected one is apparently due to the simultaneous action of two competing processes: the destruction of "biographical" defects with the transfer of phosphorus to the active state and the realization of the conditions for the reverse process - the formation of phosphorus-binding thermostable complexes in areas of growth and radiation defects.

As follows from the figures in fig. 4.21a experimental data, the largest deficiency of phosphorus is observed in dislocation crystals, which may indicate a predominant drain of interstitial atoms to dislocations. Note that some of them belong to the unstable Si^{31} isotope. Their appearance in internodes is explained by the fact that the recoil energy of silicon atoms upon emission of a γ quantum in (n,γ) nuclear reactions is sufficient to displace them from lattice sites and even beyond the unit cell (see Chapter 2). In this form, the atoms of the silicon-31 isotope should easily migrate along the crystal to localization in the region of any imperfections. The phosphorus atoms arising from them after β-decay remain in the inactive state in the region of dislocations and other microdefects, and annealing even at a temperature of 1100° C is insufficient to transfer them to the active state.

In dislocation-free crystals, various types of growth microdefects serve as sinks for interstitial atoms. In this case, as in the case of electron irradiation, microdefects are formed due to the interaction of radiation defects with growth defects, which are detected by selective etching and were absent before irradiation and annealing [37].

In addition, the deficiency of electrically active phosphorus may be due to the deposition of phosphorus atoms on radiation complexes. This case was noted in n-type silicon samples after irradiation with Si^+ ions. As noted in [38], a temperature of 1100 K is insufficient for their annealing. In [39], it was also pointed out that during the diffusion of phosphorus into silicon samples preliminarily irradiated with Ar^+ ions, the temperature of the subsequent stabilizing annealing is much higher than with ordinary ion doping.

In our case, radiation defects begin to prevail after irradiation with neutron fluxes of the order of 10^{19} cm^{-3} and higher. In accordance with the data of [32], it turned out that annealing at 1100 K is insufficient for a complete annealing of radiation defects and the transfer of the phosphorus atoms to an electrically active state.

In conclusion, we note that an annealing temperature increase above 900°C is undesirable, because then diffusion of phosphorus atoms and their enhanced capture by various structural imperfections are possible. As a result, the absolute values change and the uneven distribution of the resistivity increases [40].

4.5. Electrophysical properties of nuclear doped silicon

<u>Carrier Concentration and Mobility</u>

As noted, after annealing of radiation defects in nuclear doped silicon, stable values of resistivity, concentration and mobility of charge carriers are established. However, the question remains so far what exactly determines the stability and the magnitude of the corresponding characteristics. In particular, by analogy with ion doping, the question arises: is all phosphorus after annealing in an electrically active state. The answer to it can be obtained using an experiment in which the concentration of charge carriers is compared with the concentration of phosphorus, determined, for example, by activation analysis. In fact, the applicability of equality

$$n_i = N_d \tag{4.12}$$

which for uncompensated silicon is valid in a fairly wide temperature range [41].

It was noted above that with the accumulation of phosphorus atoms due to the (n, γ)-reaction, the concentration of the unstable phosphorus-32 isotope increases, which undergoes β^--decay with a half-life of 14.3 days. In a specially designed experiment, the number of radioactive nuclei of the P^{32} isotope was measured using a 4π counter at the Protok installation, and then the concentration of P^{31} phosphorus atoms was determined from the expression

$$\left[P^{31}\right] = \sigma_1\left[P^{32}\right]\lambda^2 t^2 Si^{30} / \left(\sigma_2\left(\lambda t - 1 + e^{-\lambda t}\right)\right) \tag{4.13}$$

which is a solution to the system of equation

$$\begin{aligned} d\left[P^{31}\right]/dt &= \varphi\sigma_1\left[Si^{30}\right]; \\ d\left[P^{32}\right]/dt &= \varphi\sigma_2\left[P^{31}\right] - \lambda\left[P^{32}\right]; \end{aligned} \tag{4.14}$$

similar to equations (2.33) and describing the accumulation of P^{31} and P^{32} isotopes, taking into account the decay of the latter. When compiling and solving system (4.14), it was assumed that the irradiation time is much longer than the half-life of Si^{31}, which is equivalent to the direct $Si^{30} \rightarrow P^{31}$ transmutation. The obtained experimental data are summarized in table 4.7. The values of resistivity calculated and measured by the four-probe method are also given there.

Table 4.7. Comparison of the concentration of charge carriers determined by the Hall effect and the concentration of phosphorus found from induced radioactivity

Determined value	Determination method	Values
Phosphorus concentration	Calculation	$(27\pm5)\cdot10^{14}$ cm^{-3}
The same	By induced radioactivity	$(28\pm4,8)\cdot10^{14}$ cm^{-3}
Carrier Concentration	Hall effect	$(26\pm5,2)\cdot10^{14}$ cm^{-3}
Resistivity	Calculation	1,6±0,3 Ohm·cm
The same	Four probe	1,7±0,12 Ohm·cm

Within the experimental errors, good agreement was obtained between the experimental and calculated values. For the experiments, we used samples cut from a p-type silicon ingot with a resistivity of 8000 Ohm·cm, obtained by the crucible-free zone melting method. The samples were irradiated with neutron fluxes of 10^{17}-$5\cdot10^{19}$ cm^{-2} and annealed in vacuum quartz ampoules at a temperature of 800°C for 1 h. It was also established that the unstable P^{32} isotope makes the main contribution to β^--radioactivity.

In [42], in a wide range of concentrations, the data obtained using the Hall effect were similarly compared with the results of measuring the phosphorus concentration from the induced γ activity that accompanies the β^--decay of Si^{31}. This confirms that annealing at 800°C provides not only sufficient completeness of removal of radiation defects, but also the transfer of phosphorus atoms to an electrically active state (provided that there are no complicating factors mentioned above).

The expression $\sigma = 1/\rho = n_i\mu_e e$ in addition to the concentration n and electron charge e, includes the carrier mobility μ_e. Obviously, the accuracy of the occurrence of resistivity values in a given interval will depend on the correct choice of the mobility value. In various literature sources, the electron mobility is given from 1300 [43] to 1900 cm²/(B.c) [44]. It can be expected that the mobility for nuclear-doped silicon will differ from the analogous values for silicon doped during the growing process.

The corresponding results of measurements of Hall mobility for a large group of samples cut from one ingot have the following form:

Neutron flux, cm^{-2}	10^{17}	$5\cdot10^{17}$	$8\cdot10^{17}$	$2\cdot10^{18}$	$5\cdot10^{18}$
μ, cm$^2\cdot$B$^{-1}\cdot$c^{-1}	1780 1700 1820	1730 1690 1750	1510 1765 1620	1810 1640 1690	1750 1530 1580

In the studied range of neutron fluxes (concentration of introduced phosphorus), no significant deviations of the carrier mobility from the average value close to 1600 cm²/(B.c.) were found.

Studies in the temperature range of 77–300 K showed that the carrier mobility in nuclear doped silicon varies with temperature as $T^{-1.7}$ [45], which differs markedly from the values equal to (-2.3) for ordinary silicon, and agrees quite well with the law predicted by the theory $\mu \sim T^{-1.5}$ (see, for example, [46]). The corresponding curves of the temperature dependence of mobility for silicon samples doped to different values of resistivity obtained by different methods are shown in Fig. 4.22.

These data indicate that during the doping process, no additional impurities appear in the bulk of the silicon samples, which could act as scattering centers, and, in addition, after annealing of radiation defects at a temperature of 800°C for an hour, there are practically no defects, effectively reducing carrier mobility. This is also confirmed by the data on fig. 4.23 data on changes in carrier concentration in the temperature range 77–300 K. The course of the curves indicates the absence of deep local levels in the forbidden zone.

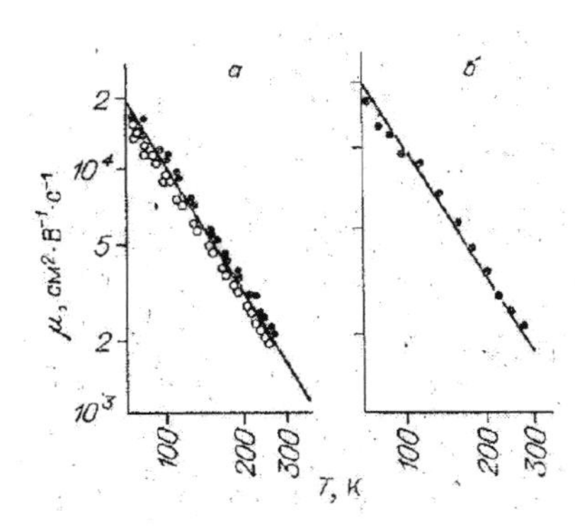

Figure. 4.22. Temperature Dependence of Electron Mobility in Nuclear Doped Silicon Obtained by Methods of Crucible-Free Zone Melting (a) and Extraction (b).

In conclusion, we note that when using the method of nuclear transmutations in silicon, obtained by the method of shell-free zone melting, it is possible to reproduce the phosphorus concentration and resistivity values fairly well in a wide range of values (see [1, 8, 23] and Fig. 4.23). However, in silicon grown by the Czochralski method and doped with a slow neutron flux in the range 10^{17}-10^{19} cm^{-3}, phosphorus atoms are not the only carrier source - after heat treatment at 700°C and above, additional centers associated with oxygen and exhibiting donor properties were found (fig. 4.24) [47]. Subsequently, data were obtained on the stability of these centers up to 1100°C [33].

Similar studies conducted by Cleland et al. [48] in a wider range of integral neutron fluxes confirmed the appearance of an additional carrier concentration in the NDS obtained by the Czochralski method, which, relatively, the greater, the lower the introduced phosphorus concentration. At phosphorus concentrations higher than 10^{16} cm^{-3}, the calculated and measured carrier concentrations coincide with an accuracy of 5% for silicon

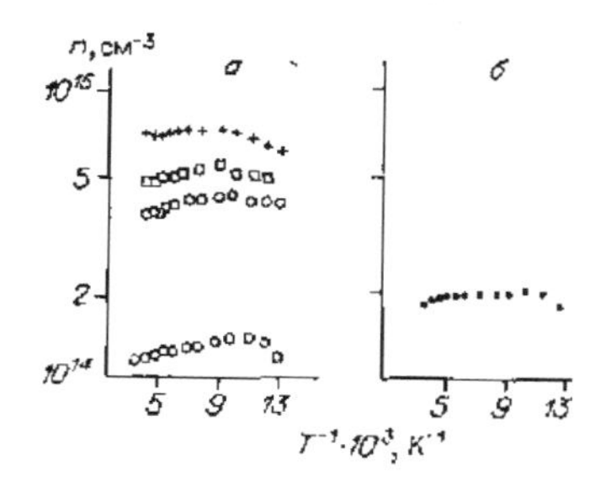

Fig. 4.23. Temperature dependence of the electron concentration in nuclear doped silicon obtained by methods of crucible-free zone melting (a) and stretching from the melt (b);

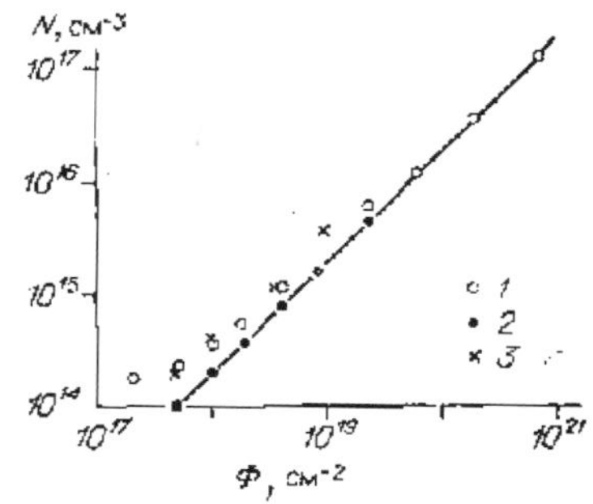

Fig. 4.24. Dependence of the concentration of donors in silicon on the neutron flux (after irradiation and annealing) [42] 1 - for silicon obtained from melt, 2 - for silicon obtained by method of zoneless melting, 3 - for silicon obtained by drawing from melt (data [47]). The solid line is the calculation.

Resistivity distribution uniformity

As noted in some works (see, for example, [9]), one of the advantages of the nuclear transmutation method is the distribution of the introduced impurity, which is close to ideally uniform. This is due to the fact that the formation of phosphorus atoms in any local volume is equally probable, since the atoms of the silicon-30 isotope in the natural mixture are distributed statistically uniformly, and the slow neutron flux is isotropic. As an example in Figure 4.25 shows the distribution of resistivity in the radial direction, measured by the current spreading method [49] on "ordinary" and nuclear doped silicon wafers. Table 4.8, taken from [50] and supplemented by more recent data, shows the results of comparative studies of the uniformity of the distribution of resistivity ρ in silicon samples obtained by different methods. N-type silicon with ρ=1000 Ohm·cm and p-type with ρ=2000 Ohm·cm and more, obtained by the

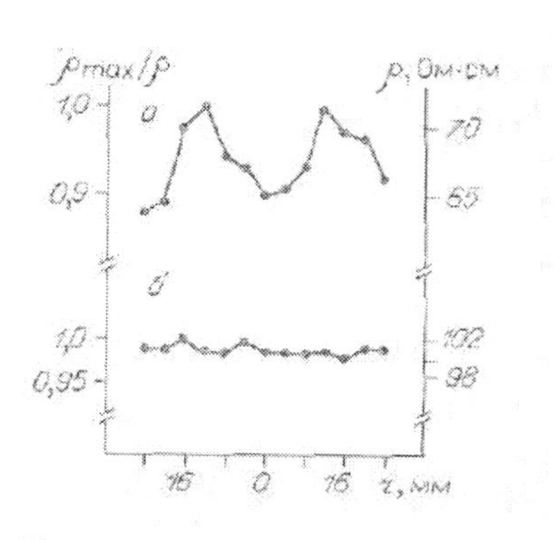

Figure 4.25. The radiation distribution of resistivity in nuclear-doped silicon (b) and in "ordinary" (a).

crucible-free zone melting, were used as the starting material for doping by the method of nuclear transmutations.

Table 4.8. Homogeneity of the distribution of resistivity in conventional and nuclear doped silicon

Production method	Resistivity, Ohm·cm	Standard deviation, %
Nuclear transmutations	30	4,5
Czochralski	30	16,0
Nuclear transmutations	70	6,0
Crucible zone melting	80	17,0
Nuclear transmutations	155	5,2
Crucible zone melting	160	19,0

Several series of measurements were carried out, each of which consisted of 150-200 measurements. Then, the standard deviation was calculated with a confidence level of 0.95. As follows from table 4.8, the standard deviations characterizing the uniformity of the distribution of ρ in the NDS are 2–3 times lower than in ordinary material.

A clear advantage of NDS is also found when comparing the ρ microdistribution. On fig. 4.26 shows the radial distributions of ρ in ordinary and in nuclear doped material 51]. In the first case, the fluctuations ρ exceed $\pm$ 10%, in the second - do not go beyond $\pm$ 5%.

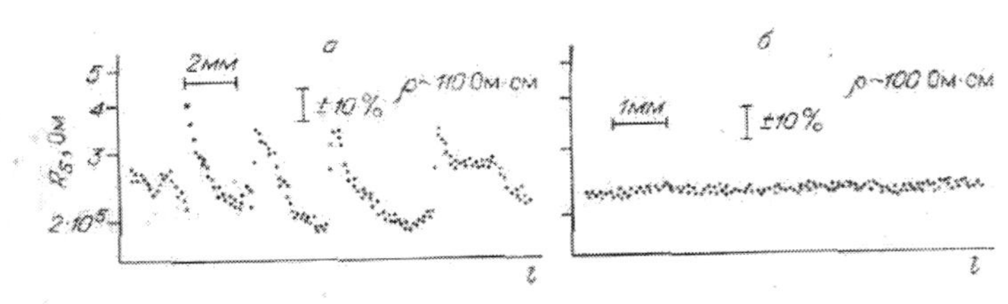

Figure 4.26. Microdistribution of resistivity in the radial direction of ordinary (a) and nuclear doped (b) silicon [45].

a - silicon is doped during the growing process; b - the same method of nuclear transmutations.

Life time of nonequilibrium charge carriers

The carrier lifetime τ in nuclear doped silicon is one of the most important characteristics of the material, which determine the limits of its practical applicability. It was noted above (see § 4.3) that after optimal annealing at temperatures of 700-800°C, it is possible to obtain fairly stable and reproducible values of resistivity, concentration and carrier mobility, and this is true for silicon irradiated with various integral neutron fluxes up to 10^{21} cm^{-2} and in different reactors with a ratio of slow to fast neutron flux from 1700 to 3-10 [48].

When trying to obtain stable values of τ, the researchers encountered a number of features of the behavior of the NDS. In particular, it turned out that in individual ingots stable values of τ are obtained at higher annealing temperatures than is necessary to obtain stable values of ρ, n_i, μ and that the recovery process τ depends on a number of factors, including "purity" and the defective structure of the initial crystals, as well as its initial resistance, exposure time at annealing temperature, furnace atmosphere, cooling rate, etc. (see [48]).

Unexpected and still unexplained is the fact of obtaining NDS with high values of τ upon irradiation with a neutron flux with a high content of fast neutrons and, therefore, under conditions of the formation of a higher concentration of radiation defects, whereas the opposite effect should be expected [48, 52].

To concretize these facts, we recall that the initial silicon obtained by the crucible-free zone melting is usually characterized by values of $\tau \sim$

Table 4.9. Lifetime of minority current carriers in radiation-doped silicon, depending on the imperfection of the structure, μs

Crystal excellence	Slow neutron flux, cm^{-2}							
	$1,4\cdot10^{17}$	$2,4\cdot10^{17}$	$5,5\cdot10^{17}$	$9,1\cdot10^{17}$	$9,2\cdot10^{17}$	$1,4\cdot10^{18}$	$2,8\cdot10^{18}$	$3,8\cdot10^{18}$
	Top end							
$5\cdot10^4$	6-10	3-5	8-10	6-10	8-10	8-12	10-12	7-12
Disfr., cl.	120-130	100-140	80-120	100-130	90-130	80-100	70-110	90-120
	Bottom end							
$5\cdot10^4$	3-6	2-4	7-10	2-3	8-10	8-12	8-15	10-15
Disfr., cl.	30-80	50-70	30-50	60-80	40-80	30-50	40-50	40-70

Note: $5\cdot10^4$ cm^{-2} – dislocation density; disfr. – dislocation free; cl. – clusterless.

1–3 ms [53]. Immediately after irradiation, the values of τ are so small that they are practically impossible to measure. During high-temperature processing, as the defects are annealed, the values of τ are restored (Fig. 4.27). In this case, the final values of τ substantially depend on the perfection of the crystals and weakly on the neutron flux (Table 4.9). However, complete restoration of the initial values of τ does not occur.

This effect is most pronounced in silicon single crystals with a dislocation density of ~(2-5)·10^4 cm^{-2} [9, 22], and for dislocation free silicon, the minimum values are characteristic of crystals with growth microdefects, and there is a correlation between the concentration of microdefects and the degree of degradation carrier lifetime [54] (Fig. 4.28).

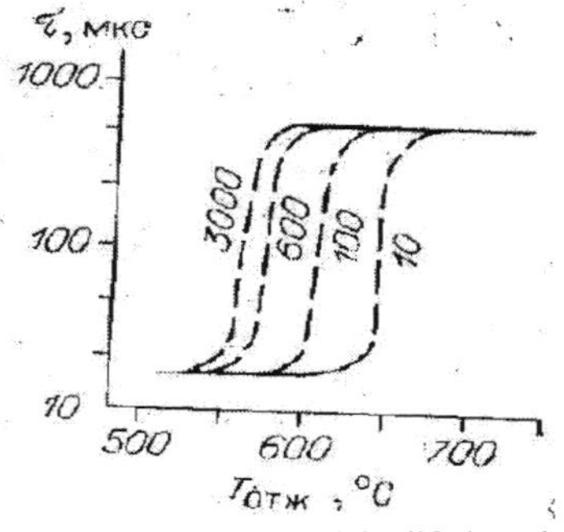

Figure 4.27 Recovery of the lifetime during isochronous annealing of silicon irradiated with reactor neutrons with different ratios of thermal and fast neutrons [26].

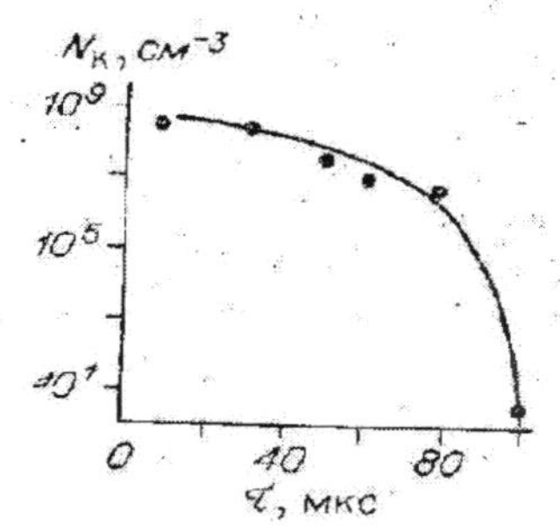

Figure 4.28. The relationship between the density of defects and the lifetime of charge carriers in nuclear doped silicon

A typical distribution of microdefects on the (111) plane and the character of the distribution of the lifetime measured by the modulation method of conductivity in a point contact are shown in Fig. 4.29. The largest changes in τ are observed in the region of the maximum concentration of growth microdefects (bright fields). In places free from defects (dark fields), the life time in relation to the original one has changed to a lesser extent. The value of τ changed more significantly in crystals, where microdefects were etched in the form of knolls, which are probably precipitates of metallic impurities.

At present, there are still insufficient experimental facts to describe in detail the mechanisms of degradation of the carrier lifetimes in the NDS.

Therefore, methods for reliable control of this parameter during irradiation and subsequent heat treatment are not yet clear, although attempts to solve this problem are known. Thus, according to [48], the best results can be achieved by annealing the NDS obtained by the crucibleless zone melting for an hour at 1000°C in a getter atmosphere. The cooling rate is chosen so that there are no residual quenching phenomena. In this regard, for annealing defects, it is recommended to cool the NDS from 1000 to 650°C at a speed of 20 deg/h, then it is necessary to ensure rapid cooling from 650°C to room temperature and thereby prevent the formation of oxygen complexes in the region of 450°C. Note that the last recommendation does not work in the case of nuclear doping of silicon ingots with a diameter of 100-150 mm or more, since due to insufficient thermal conductivity it is impossible to provide the required cooling rate. Therefore, measurements of the electrophysical parameters of such ingots are carried out on control washers up to 10 mm thick, cut from the initial ingots to doping.

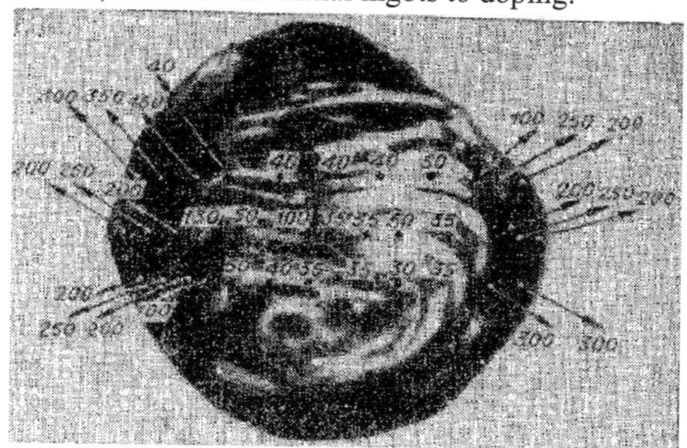

Figure 4.29. Distribution of lifetimes and growth defects at the end face of a silicon ingot doped with nuclear transmutations

A certain optimism is inspired by a clearly observed tendency to increase the lifetime with increasing purity of the starting material (for example, with respect to carbon). This is apparently due to the fact that with a decrease in residual impurities, the number of centers of microdefect nucleation decreases and, accordingly, the probability of the appearance of new recombination centers decreases.

The possibility of recovering τ in the NDS upon annealing to almost initial values is described below.

Structural defects

Real single crystals always contain one or another number of structural imperfections. These include: point defects, including impurity atoms and their complexes; linear defects - dislocations; two-dimensional defects such as dislocation loops and stacking faults, surface and boundaries; bulk imperfections (pores, particles of foreign phases, etc.).

In chapter 3 shows that radiation defects actively interact with structural imperfections both at the irradiation stage and during annealing. In this case, the sizes, density, and electrical activity of birth defects vary [55]. As a result, such characteristics of the NDS, such as, for example, recombination ones, will differ from the characteristics of the starting material. In addition, the consequences of this interaction, as will be shown below, affect the homogeneity of the distribution and the accuracy of getting into the specified limits of the values of resistivity.

From the point of view of changes in structural defects under the action of radiation treatment, single-crystal silicon can be divided into three groups. The first group includes crystals characterized by the presence of a certain set of growth defects, the density of which under the influence of radiation treatment can change. Such crystals are quite easily detected and rejected even at the stage of obtaining the starting material. The second and third groups include silicon single crystals, in which defects are not detected prior to irradiation. They differ in response to radiation treatment: in some single crystals, after irradiation and annealing, an uncontrolled appearance of structural defects is observed. In order to prevent such a process, it is necessary to know the mechanisms of interaction of radiation and growth defects. These include:

1. The decomposition of precipitates and the resorption of atoms of impurities localized in the region of growth defects. When this mechanism is realized, the density of dislocations and two-dimensional defects in dislocation-free crystals changes as follows:

	Crucible zone melting method				
Sample Number	1	2	3	4	5
Neutron flux, cm^{-2}	$2,3\cdot10^{17}$	$2,2\cdot10^{17}$	$6,2\cdot10^{17}$	$1,0\cdot10^{18}$	$2,5\cdot10^{17}$
	Etch pit density, cm^{-2}				
Original	$9,8\cdot10^{3}$	$6,3\cdot10^{3}$	$8,3\cdot10^{3}$	$2,5\cdot10^{4}$	With. disl.

After irradiation	$2,4\cdot 10^{4}$	$1,3\cdot 10^{4}$	$1,2\cdot 10^{4}$	$3,0\cdot 10^{4}$	»
After subsequent annealing	$2,3\cdot 10^{4}$	$1,3\cdot 10^{4}$	$1,2\cdot 10^{4}$	$3,0\cdot 10^{4}$	»
			Czochralski method		
Sample Number	1	2	3	4	5
Neutron flux, cm^{-2}	$2,3\cdot 10^{17}$	$4,0\cdot 10^{17}$	$2,5\cdot 10^{17}$	$3,1\cdot 10^{18}$	$2,5\cdot 10^{17}$
			Etch pit density, cm^{-2}		
Original	$5,5\cdot 10^{3}$	$6,0\cdot 10^{3}$	$1,3\cdot 10^{3}$	$7,2\cdot 10^{4}$	With. disl.
After irradiation	$9,1\cdot 10^{4}$	$7,4\cdot 10^{4}$	$1,3\cdot 10^{4}$	$1,0\cdot 10^{4}$	»
After subsequent annealing	$2,0\cdot 10^{4}$	$4,3\cdot 10^{4}$	$5,2\cdot 10^{4}$	$2,4\cdot 10^{4}$	»

In addition, in the case of dislocation free crystals grown in an argon atmosphere with hydrogen, after radiation treatment, anomalous etching sites appear, similar to those noted in [36]. The dimensions of these defects reach several hundred microns, and it is not possible to fabricate operable *p-n-* junctions from silicon with such defects. A characteristic feature of two-dimensional defects is their distribution, reflecting mainly the crystallization front (Fig. 4.30).

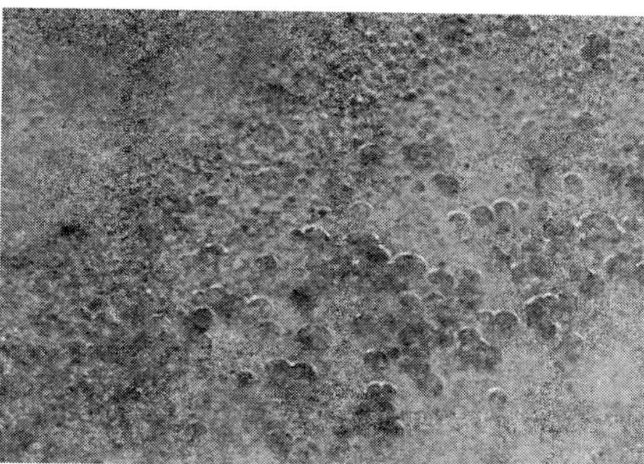

Figure 4.30. The character of the distribution of microdefects at the end face of a silicon ingot doped by the method of nuclear transmutations.

2. The formation in the region of growth defects of precipitates from radiation defects, atoms of any impurity or their complexes. This mechanism was studied in sufficient detail in [55], where it was shown that dislocations are effective sinks for radiation defects. So, in silicon obtained by the Czochralski method, it was found that after irradiation

and annealing, some of the oxygen is localized in the region of dislocations. Oxygen atoms are likely to fall into the dislocation region during annealing as a result of runoff in the form of A centers. According to [56], localized oxygen effectively prevents selective etching, which manifests itself in a change in the shape of etching pits and their density (see paragraph 1).

In dislocation-free crystals, as a result of the drain of radiation defects onto growth microdefects during irradiation and annealing, the size of the latter becomes higher than critical, and such defects begin to be easily detected. In addition, during the deposition of particles of a foreign phase, formation of dislocation loops around them from silicon atoms displaced from the bulk of the second phase is quite likely.

3. Radiation-stimulated coalescence of growth defects of small sizes, possible during irradiation and subsequent annealing

4. The formation of defects of radiation origin, which can be detected by selective etching [57]. Similar changes in structural defects, according to [37], are also observed in silicon crystals irradiated with fast electrons, which create only point defects.

In each case, the interaction mechanism is determined by the initial state of the defects in the crystal and the presence of an impurity both in the bulk and near the defects.

4.6. Effect of annealing medium on the properties of nuclear doped silicon

A thorough analysis of the results of studies of the electrophysical properties of representative batches of nuclear doped silicon (NDS) in the form of full-scale ingots obtained in the early stages of the development of pilot industrial production confirmed the main advantage of nuclear doping in terms of the specific electrical resistance parameter - the accuracy of falling into specified limits was at the level of 90 -95%, the scatter at the micro- and macroscale did not exceed 5-7%. However, in terms of the parameter, the lifetime of minority charge carriers (NCCs) of NDS was significantly inferior to silicon doped during the growth, both in absolute value and in scatter from ingot to ingot and within one ingot. Since the degradation of the lifetime of minority charge carriers is associated with the presence of recombination centers in the crystal, primarily in the form of structural defects, impurity atoms with deep levels and their complexes, the question naturally arose of the completeness of annealing of defects and unauthorized contamination of ingots by rapidly diffusing impurities. In [58–61], purposeful studies of the influence of the annealing medium and the irradiation conditions in a nuclear reactor were carried out, as well as the possibility of eliminating negative factors that effectively reduce the lifetime. In the first series of experiments, studies were made of the influence of the annealing medium. Single crystals obtained by the crucibleless zone melting were irradiated in a WWR-t reactor with a thermal: fast neutron ratio of 20: 1 thermal neutron fluence of $\sim 1 \cdot 10^{17}$ cm^{-2}, which corresponded to doping with a nominal value of ρ ~180 Ohm·sec.

After irradiation, cylindrical ingots were cut along the generatrix into four equal parts. Each part of the ingot was annealed at a temperature of 850°C for 4 hours in one of four media: dynamic vacuum with oil-free pumping (10^{-6} torr), air, technical argon duct, and chlorine-containing atmosphere (CCA). A mixture of purified oxygen and argon saturated with carbon tetrachloride vapors (the molar concentration of the latter was 1%) was used as CCA. Diffusion tubes were made of highly pure quartz, and upon annealing in vacuum and in air, the samples were placed in closed silicon containers. Before irradiation and annealing, the samples were subjected to chemical etching and washing in peroxide-ammonia and acid-peroxide solutions.

After annealing, the samples were cut across into two equal parts, on the external and "internal" ends of which the resistivity was measured by the four-probe method, the lifetime of minority charge carriers by the method of conduction modulation in a point contact, and the microstructure was studied after selective etching in a hot Sirtl etch.

The results of measuring the resistivity are shown in the table 10, where ρ_o and ρ_i are the average values of resistivity, measured respectively at the outer and inner ends, ρ_o^{av} and ρ_i^{av} are the average values of resistivity at each end for four annealing media.

It can be seen from the above data that the difference in the values of ρ_o and ρ_i for the same ingot after annealing in different media is no more than 5%, which does not exceed the expected calculated values. This result, as well as the absence of a correlation between the scatter ρ along the ends and in the volume and the heat treatment medium, show that the resistivity and its scatter are determined only by the uniformity of irradiation and the scatter of resistivity in the initial samples and are independent of the annealing atmosphere.

Table 4.10. Resistivity measured at the outer and inner ends of the NDS samples cut from one ingot and annealed in different media

Ingot number	Measured value	Annealing medium			
		CCA	Vacuum	Argon	Air
1	ρ_o, *Ohm*·s	236	235	243	248
	ρ_o/ρ_o^{av}	0.98	0.98	1.01	1.03
	ρ_i, *Ohm*·s	244	247	253	258
	ρ_i/ρ_i^{av}	0.97	0.99	1.01	1.03
2	ρ_o, *Ohm*·s	187	190	188	195
	ρ_o/ρ_o^{av}	0.98	1.00	0.99	1.03
	ρ_i, *Ohm*·s	188	189	186	195
	ρ_i/ρ_i^{av}	0.99	1.00	0.98	1.03
3	ρ_o, *Ohm*·s	185	181	186	187
	ρ_o/ρ_o^{av}	1.00	0.98	1.01	1.01
	ρ_i, *Ohm*·s	183	182	184	184
	ρ_i/ρ_i^{av}	1.00	0.99	1.00	1.01
4	ρ_o, *Ohm*·s	170	173	184	183
	ρ_o/ρ_o^{av}	0.96	0.97	1.03	1.03
	ρ_i, *Ohm*·s	175	172	181	189
	ρ_i/ρ_i^{av}	0.98	0.96	1.01	1.05
5	ρ_o, *Ohm*·s	186	190	190	190
	ρ_o/ρ_o^{av}	0.98	1.01	1.01	1.01
	ρ_i, *Ohm*·s	190	193	189	194
	ρ_i/ρ_i^{av}	0.99	1.01	0.99	1.01

Significantly large differences were observed in the results of measuring the NNC lifetime for silicon samples annealed in various media. The results of measurements of the NCC lifetime are presented in Fig. 4.31. As can be seen from the figure, the values of the NCC lifetime increase in the series: technical argon - vacuum - air - CCA. In this case, the values of the NCC lifetime were lower on the outer ends of the samples than on the inner ones obtained after cutting the annealed samples.

The data presented convincingly indicate diffusion during annealing from the external environment of uncontrolled impurities, contributing to the emergence of recombination centers. The greatest pollution occurs during annealing in technical argon and vacuum. During annealing in CCA and in air, uncontrolled impurities are apparently bound by chlorine and oxygen, as a result of which their penetration into the samples is partially prevented. The above considerations are confirmed by the results of studies of the microstructure of the samples. After annealing the samples in technical argon in the entire volume of the samples and after annealing in air in the surface layer with a thickness of up to 5 mm, microdefects (MD) with a concentration of 10^6-10^7 cm^{-3} were found that had a uniform or spiral distribution. At the same time, after annealing in vacuum and in CCA, only in two crystals a small amount of MD appeared in the near surface layer.

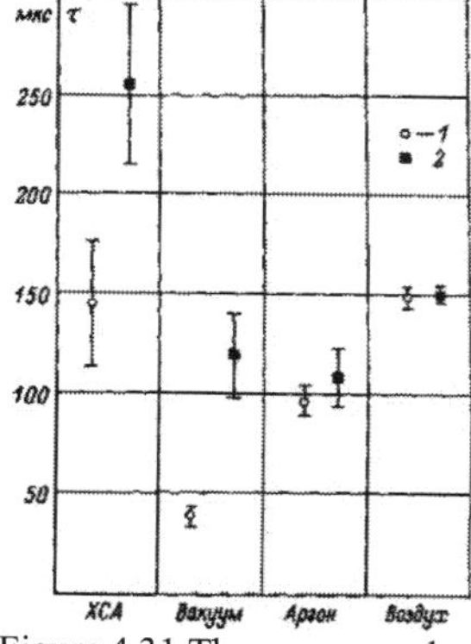

Figure 4.31 The average value and variance of the lifetime of minority charge carriers after heat treatment of irradiated crystals in various media. 1 - outer end, 2 - inner end

Estimates based on the size of the samples and the duration of the heat treatment showed that penetration of the recombination-active impurity to the entire depth of the samples can be achieved at a sufficiently high diffusion coefficient. Given this factor, it can be assumed that diffusion is facilitated in irradiated crystals due to the presence of excess intrinsic point defects. During annealing of the crystals in CCA, their pollution from the outside by uncontrolled impurities is prevented and the concentration of excess interstitial silicon atoms is reduced, which, along with impurity atoms, are responsible for the formation of recombination centers and MDs, therefore, their recombination characteristics and structural perfection are due to impurity atoms and growth contained in the initial crystals MD; as a result, the obtained values of the lifetime are close to the initial ones, but at the same time they differ significantly from crystal to crystal (large dispersion in Figure 4.31).

Thus, it follows from the obtained results that the annealing medium does not affect the resistivity of the NDS at least within a spread of + 5%, but causes a significant difference in the values of the lifetime and microstructure, which is associated with the penetration of impurity atoms from the external medium into the volume of doped crystals , and which contribute to the emergence of recombination centers as well as the decoration of hidden growth microdefects. Single crystals annealed in CCA possess the longest life time of minority charge carriers and structural perfection. It can be assumed that the surface of the ingots is the main source of contamination of the volume of the single crystal. This assumption is supported by evidence that silicon adsorbs various elements, including metals, very effectively from the medium in contact with it.

The next step was to study the low-temperature diffusion of certain impurities adsorbed by the silicon surface upon irradiation in a nuclear reactor and from special solutions before heat treatment of the irradiated samples, as well as the effect of the diffusant on the electrophysical properties of nuclear doped silicon.

For experiments, we used samples of dislocation-free single crystal silicon grown by the method of crucible-free zone melting, *n*-type conductivity, orientation (111) with a specific electrical resistance of ~ 400 Ohm·cm and a lifetime of nonequilibrium charge carriers of 300-400 μs. The samples were irradiated with a thermal neutron fluence (Φ) = $1\cdot10^{17}$-$1\cdot10^{18}$ n/cm^2 in contact with water used to cool the samples during irradiation, which contains Na, Cr, Fe, Au, Sb, La, Zr and other elements [see Further].

After irradiation, the samples were washed in an acid-peroxide solution (APS) of the composition H_2O_2: HNO_3: H_2O. Radioactive gold control washing using this solution turned out to be quite effective:

Washing time, min	0	10	20	30
The amount of remaining gold, wt. %	100	1,88	0,70	0,52

Within 20-30 minutes of washing, gold is almost completely removed from the surface of the sample. Therefore, in all cases, washing in APS was used to remove elements that were secondarily adsorbed from the solution.

The heat treatment of the irradiated and control (non-irradiated) samples was carried out in air in a muffle furnace, in air in a quartz tube of an diffusion furnace, and in a chlorine-containing atmosphere (CCA). The electrical resistivity (ρ) was measured by the four-probe method and the lifetime τ by the modulation of conductivity in a point contact. The data are summarized in table 4.11.

As noted above, and in this experiment, there is no significant effect of the heat treatment medium on the value of electrical resistivity and its scatter. A very sensitive characteristic to the "purity" of the heat-treatment medium was τ. The lowest values of τ were in samples of NDS annealed in a muffle furnace, and the highest were in samples heat-treated in CCA. Thus, these data clearly indicate the diffusion of contaminants from the outside - from the internal atmosphere of a heated furnace.

Table 4.11. Average values of electrical resistivity and lifetime of nonequilibrium charge carriers in samples of nuclear doped silicon annealed in different media

Φ, neutron/cm²	ρ, Ohm·cm			τ, μs		
	1	2	3	1	2	3
10^{17}	82	80	80	30	70	200
10^{18}	16	15	14	10	30	150
0*	160	151	149	70	130	300

Note. Φ - fluence of thermal neutrons. 1 - annealing in air in a muffle furnace, 2 - annealing in a quartz ampoule, 3 - annealing in a chlorine-containing atmosphere.
*Control unirradiated sample.

The conclusion that the decrease in τ is due to the diffusion of impurities from the surface of annealed ingots is confirmed in the following experiment. Two groups of samples were irradiated in the "wet" and "dry" channels in direct contact with the irradiation medium. The third group of samples was irradiated in a sealed quartz ampoule. After irradiation, all samples were washed once only in distilled water. Thus, the elements chemisorbed during irradiation remained on the surface of the samples. The results of measurements of τ after annealing are given below:

Exposure conditions	«wet» channel	«dry» channel	sealed ampoule
Ratioradioactivity, rel. units	10	1	0.1
τ, μs	5	80-100	150

Minimum values of τ are characteristic of samples irradiated in contact with water. The values of τ in the samples irradiated in the "dry" channel are noticeably higher. Qualitatively, the composition of the adsorbed elements in both cases is almost the same, but they differ quantitatively, which ultimately manifested itself on the value of τ. The highest values of τ were in the samples irradiated in a hermetic ampoule. Since in this experiment the heat treatment of the irradiated samples was carried out under identical conditions, a decrease in τ can only be attributed to the diffusion of contaminants from the adsorbed layer.

In order to find out the distribution pattern of some impurities in the crystal volume and, accordingly, the distribution pattern τ, we determined the diffusion coefficients for some of them after annealing at a temperature of 820°C. Figure 4.32 shows the distribution profiles after heat treatment of the concentrations of the donor impurity Sb, the rapidly diffusing impurity Au and the neutral impurity La adsorbed by the surface of the silicon samples during irradiation in the "wet" channel. The distribution of the impurity over the depth of the samples was investigated using activation analysis by the method of stripping. According to Fig. 4.32 determined the diffusion coefficients of these elements. It should be noted that the graph of the specific activity *ln*A(x,t) versus x^2 (x is the layer thickness) is a broken line with different slopes. The section with the maximum slope characterizes diffusion in the surface layer with a thickness of up to 2-4 microns. A flat section corresponds to diffusion in volume.

The calculated surface and volume diffusion coefficients for various elements are:

Radioactive element	^{122}Sb	^{124}Sb	^{140}La	^{198}Au
$D_n \cdot 10^{14}$, cm^2/s	6,9	7,1	1,2	15
$D_o \cdot 10^{14}$, cm^2/s	2,8	2,4	1,3	49

As expected, the diffusion coefficient of gold is approximately an order of magnitude higher than the diffusion coefficient of the donor impurity Sb.

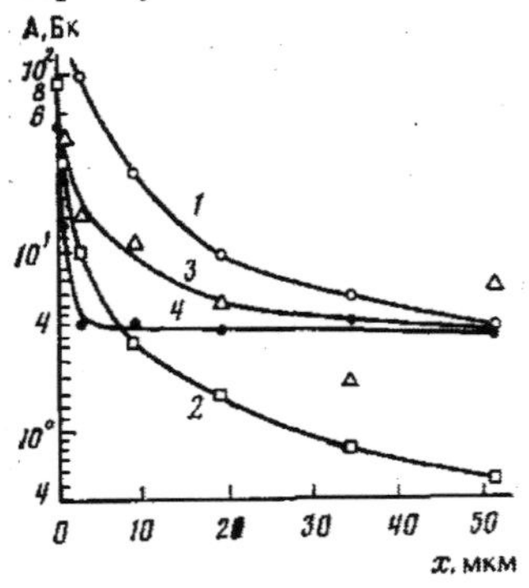

Figure 4.32 Distribution of radioactive elements ^{122}Sb (*1*), ^{124}Sb (*2*), ^{140}La (*3*), ^{198}Au (*4*) in the volume of nuclear doped silicon samples irradiated with a neutron fluence of $5 \cdot 10^{18}$ neutrons/cm^2.

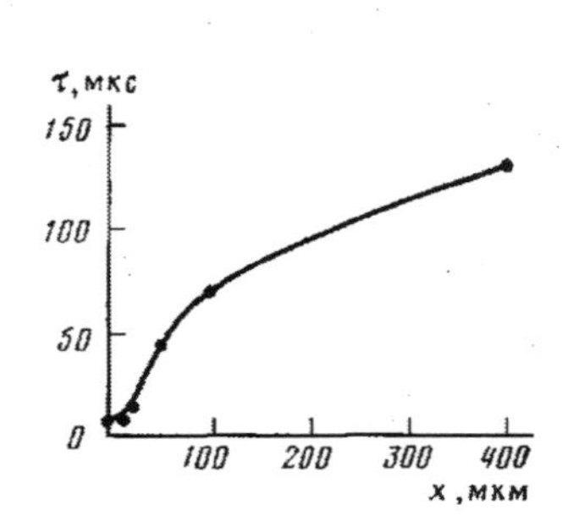

Figure 4.33 Distributions of the lifetime of nonequilibrium charge carriers in a sample of nuclear doped silicon with pre-adsorbed gold (irradiation with a neutron fluence of $2.5 \cdot 10^{18}$ neutrons/cm^2, annealing at 820°C, 2 h.

To reveal the effect of gold diffusion on the drop in τ, an additional experiment was performed in the NDS, in which a radioactive gold isotope was deposited on their surface before irradiation and before heat treatment of the irradiated and control samples. In all cases when gold was present on the samples, the life time measured after heat treatment was significantly lower than the life time in the control samples. On Figure 4.33 shows the change in τ over the volume of the sample, on the surface of which gold was deposited before heat treatment after irradiation. As layers with a higher gold content are removed, the lifetime increases, reaching a certain stationary value. It should also be noted that radiation defects stimulate the diffusion of gold: the diffusion coefficient of gold determined in this experiment at 820°C is an order of magnitude higher for irradiated samples relative to unirradiated ones.

Thus, lower lifetimes of nonequilibrium charge carriers in nuclear doped silicon are caused by contamination during heat treatment by rapidly diffusing impurities of ingots from the adsorbed layer.

It can be assumed that the impurities that penetrated during annealing into the silicon volume upon creation of appropriate conditions conducive to rediffusion will leave the ingot volume the more actively, the higher the diffusion coefficient. This is confirmed by an experiment in which irradiated silicon ingots that underwent heat treatment in air and in which τ decreased to values of 30–50 μs were subjected to repeated heat treatment in CCA, as a result of which the lifetime of nonequilibrium charge carriers increased to values of 200–220 μs . A similar effect of an increase in τ is observed after heat treatment in CCA and in "ordinary" silicon ingots, but having small initial values of τ.

In this connection, the question remains open of experimental confirmation of the possibility of back diffusion of a rapidly diffusing impurity during the heat treatment of silicon samples in CCA near 800°C.

Next, we present the results of studies of the behavior during the heat treatment of iron and gold, previously introduced into single-crystal silicon samples from an adsorbed surface layer. The initial samples were cut from a dislocation-free single-crystal *n*-type silicon ingot grown by the crucible-free zone melting method. The electrical resistivity and lifetime of nonequilibrium charge carriers (NCC) were 400 Ohm·cm and 900-1000 μs, respectively. Three groups of samples were examined. In all three groups, gold and iron were planted in the form of ^{195}Au and ^{59}Fe radioactive isotopes on one of the ends polished and washed in an acid-peroxide solution. Samples of the first group were subjected to heat treatment at 820°C for 2 h in air. The second group of samples after heat treatment in air was annealed in CCA under the same conditions. After planting Au and Fe, the third group was heat treated only in CCA.

The results of measuring the total radioactivity of the samples after various heat treatments are shown in table 4.12. It can be seen that the activity of the samples treated at elevated temperatures in air decreases slightly, the impurity mainly diffused into the bulk of the sample. The radioactivity of the samples of the second series, additionally processed in CCA, is reduced to 60%. Apparently, some of the atoms of the rapidly diffusing impurity penetrated into the volume during annealing in air and then, after heat treatment in the CCA, left the bulk of the samples to the surface and were removed in the form of gaseous chlorides by the gas flow from the working volume of the furnace. The radioactivity of the samples of the third series (processed only in CCA) decreased by 5-6 times for samples with a polished surface and 10 times for samples with a polished surface.

Table 4.12. Residual radioactivity of silicon (A) samples (^{195}Au isotope) after various stages of heat treatment.

Annealing conditions	A, Bk (cm^2)				
	after adsorption ^{195}Au	After annealing in air 820°C, 2 h	After annealing in CCA 820°C, 2 h	after 1st grinding	after 2nd grinding
1st series: annealing in air in a quartz ampoule, 820°C, 2 h	160,0	156,3		91,4	52,3
2nd series: annealing in air in a quartz ampoule, 820°C, 2 h	300,0	285,2	174,0	25,3	17,8
CCA annealing, 820°C, 2 h	200,0	195,0	105,7	9,8	7,1
3rd series: annealing in CCA 820°C, 2 h	156.3 33,6 (polished surface)		23,7 3,73	2,9 1,32	1,9 1,10

Note. CCA - Chlorine-containing atmosphere.

In addition to determining the total radioactivity, the concentration profile of the distribution of elements in samples of all three series was studied by the method of removing layers. The gold distribution profile is shown in Fig. 4.34a, iron - in Fig. 4.34b. It is noteworthy that curves 2 and 3 in Fig. 4.34b practically coincide, in fig. 4.34a, curve 2 is located between curves 1 and 3. This suggests that in the first case, the diffused iron, after heat treatment in the CCA, almost completely leaves the samples. The radioactivity of samples of the 2nd and 3rd series at a depth of 15 μm decreases almost the same - almost 2 orders of magnitude.

Gold behaves somewhat differently. In the samples of the second series (curve 2, Fig. 4.34a), activity at a depth of 15 μm decreases by an order of magnitude, but remains quite noticeable. In the samples of the third series processed immediately in CCA, gold penetrates into the sample volume in very small quantities (curve 3, Fig. 4.34a). The concentration of radioactive gold in these samples at a depth of 15 μm is approximately 2 orders of magnitude lower than in samples of the first series.

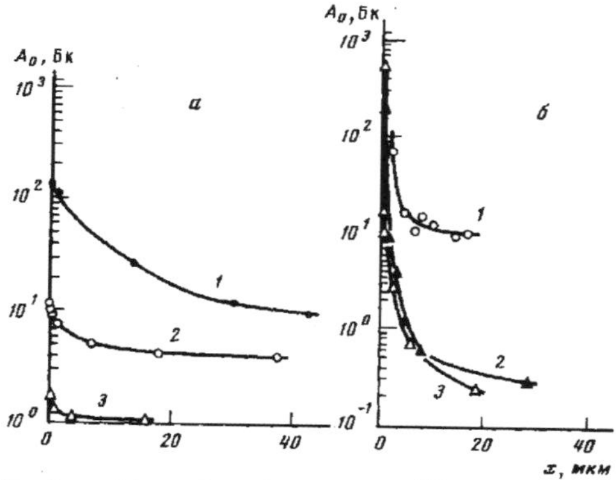

Figure 4.34. Distribution of the radioactive elements ^{195}Au (a), ^{59}Fe (b) in the volume of silicon samples after adsorption on the surface of these elements and subsequent heat treatment (1-3 correspond to the annealing conditions of the 1st – 3rd series of samples in the table)

In Figure 4.35 shows the distribution profile of the values of the lifetime of the NCC (τ) over the thickness of volumetric samples, gold was planted on one of its ends. In the surface region of the samples of all three series, τ is lower in value than the initial values. However, the degree of degradation τ depends on the type of heat treatment. The greatest changes in τ are observed in the samples of the first series (Fig. 4.35, a), slightly smaller in samples of the third series (Fig. 4.35, c). For the same samples, approximately from the same depths, a sharp increase in the values of τ begins. The distribution profile of τ in the samples of the second series is more diffuse (Fig. 4.35, b). Curve 2 shifts toward large values of τ and approaches curve 1; the course of curve 2 is similar for samples of the third series. In both cases, the effect of the chlorine-containing atmosphere on the concentration of gold in the near-surface region of the "golden" end is likely to affect. Let us single out another difference: in the samples of the first and second series, as the layers are removed, τ, in the best case, approaches 300-400 μs, and in the samples of the third series, after taking 3-4 mm, τ begins to approach the initial values and reaches 1000 μs.

Thus, the experimental data convincingly indicate the possibility of leakage in silicon under certain conditions of back diffusion of rapidly diffusing impurities. Obviously, the diffusion of rapidly diffusing impurities from the bulk of the crystal is stimulated by chemical processes on the surface in the gas solid system. For example, etching the surface with chlorine activates the formation of vacancies on the surface, which leads to the influx of interstitial silicon atoms from it, in our opinion, also atoms of rapidly diffusing impurities, in this case iron and gold. In this case, it is possible that not all impurity atoms can migrate to the surface at the same rate. Due to some features in the diffusion mechanisms, a certain fraction of

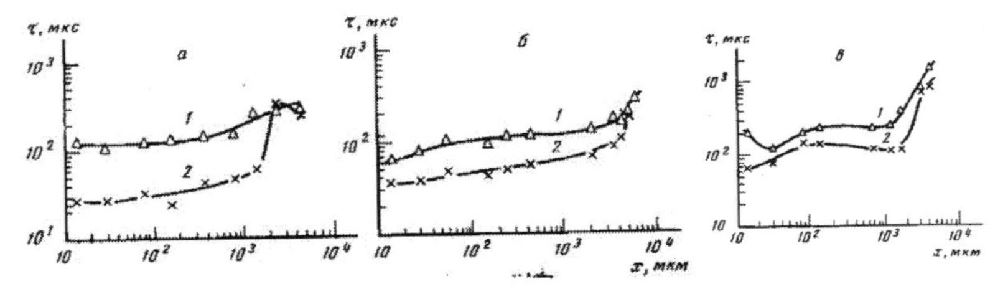

Figure 4.35. Distribution of τ in silicon samples for the clean end face of the ingot (1) and for the end face with adsorbed gold (2) (a-c correspond to the annealing conditions of the 1st-3rd series of samples in the table)

dissolved gold atoms, including those localized in the region of growth microdefects, remains in the crystal (Fig. 4.34, a, curve 2). In turn, iron atoms almost completely come to the surface (Fig. 4.34, b, curves 2, 3), where they easily form a volatile compound and are removed from the working medium.

Next, a statistical analysis of the reproducibility of the above features of annealing of nuclear doped silicon in different media under conditions of "mass" production was carried out. For research, 200 ingots that went through a complete technological cycle, but were annealed in different media: in vacuum, in air, and in CCA, were selected from the stream.

On the fig. 4.36, a presents histograms of the relative deviations of the electrical resistivity from a given doping rating (DR). Deviations of the mean value from the DR were 0, -1, and + 2% for crystals annealed in air, vacuum, and CCA. The best coincidence of the mean value with DR in the case of annealing in air is due to the fact that control ingots were annealed in air before choosing a radiation dose. During annealing in different media, as a rule, a different number of uncontrolled electrically active centers with a low level of occurrence are introduced, and which, although insignificant, affect this parameter. Therefore, using preliminary annealing in an appropriate medium and adjusting the radiation dose, it is not difficult to ensure the most accurate coincidence of the average and nominal values of resistivity.

Histograms of the distribution of the relative scatter of the resistivity at the ends of the ingots are shown in Fig. 4.36, b. It can be seen that 86% of single crystals after annealing in air, 72% in vacuum, and 83% in CCA have scatter at the ends of less than 5%. The scatter in the NDS is determined by the uniformity of the distribution of resistivity over the initial ingot, the irradiation conditions (uniformity of the neutron flux and doping factor) and annealing parameters. In this case, the first two factors were identical. This suggests that the uniformity of the distribution of introduced uncontrolled centers that change the resistivity is at the same level for annealing in air and in CCA and decreases during annealing in vacuum.

The histograms of the distribution of the lifetime of nonequilibrium charge carriers τ are shown in p. 4.37, a. For annealing in air, in vacuum, and in CCA, the average lifetimes in NDS ingots were 140, 150, and 340 μs; most of them (at least 80%) have a value of τ in the ranges 70–190, 90–210, and 230– 450 μs. Note that the lifetime after annealing in CCA is determined by the lifetime in the initial single crystals, whereas in the case of annealing in air or in vacuum it is practically independent of the initial values.

After annealing of radiation defects in air or in vacuum, part of the ingots does not meet the technical requirements due to low values of τ. The possibility of reducing τ in such crystals by additional heat treatment in CCA was investigated.

In Fig. 4.37, b shows the values of τ before (dashed line) and after (solid line) annealing in CCA of the order of 30 ingots. It can be seen that the additional heat treatment in the CCA allows one to increase τ by an average of more than 100 μs.

Thus, the results of detailed studies of the influence of the annealing medium on the electrophysical parameters of NDS given in this section convincingly indicate the promise of using heat treatments in CCA in order to obtain single-crystal silicon with extremely high parameters. With sufficiently accurate observance of the technological regimes of nuclear doping, it is possible to achieve a very high reproducibility of the required parameters.

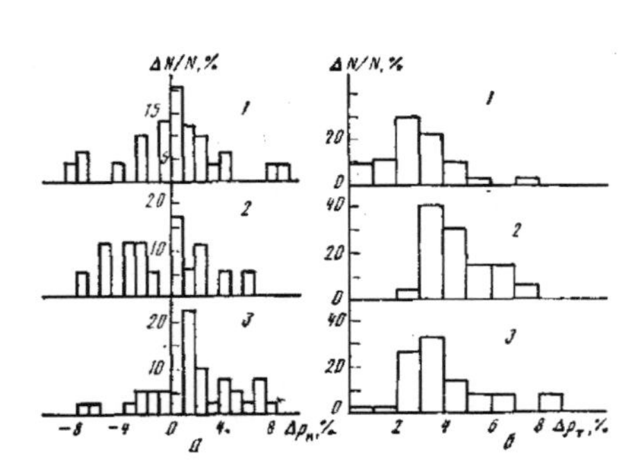

Figure 4.36. Histograms of the distribution of the relative deviation of the resistivity from the doping value (a) and the relative end-face scatter of the resistivity (b) of silicon single crystals annealed in air (1), in vacuum (2), in a chlorine-containing atmosphere.

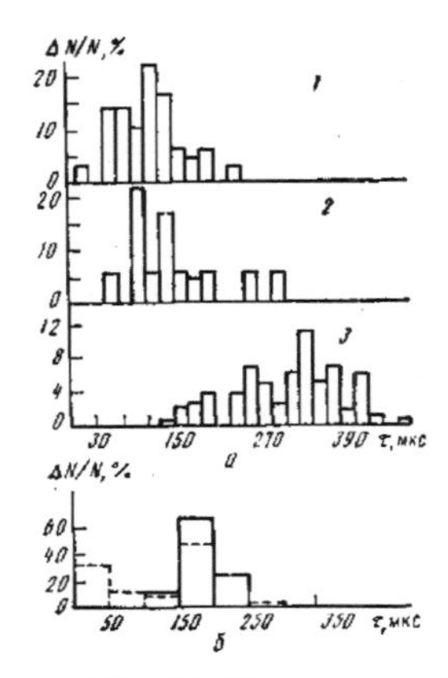

Figure 4.37. (a) Histograms of the distribution of the lifetime of minority charge carriers in silicon single crystals annealed in air (1), in vacuum (2), in a chlorine-containing atmosphere (3), (b) a histogram of the distribution of the lifetime in the NDS of the initial values (dashed line) and after (solid line) annealing in a chlorine-containing atmosphere.

4.7. The influence of the prehistory of the material and the parameters of the neutron flux on the electrophysical properties of nuclear doped silicon

In [9], the factors affecting the quality of the NDS are identified and briefly analyzed. These include: the uncontrolled formation of thermal donors or acceptors, the capture of current carriers and dopant atoms by both birth defects and those formed during radiation processing.

The uniformity of the distribution of resistivity in the NDS depends on the fluctuation of residual impurities in the initial silicon, as well as on the homogeneity of the neutron flux in the irradiation zone. In addition, the uniformity of the distribution also depends on the volume effect due to the weakening of the flux due to neutron capture as they diffuse into the sample volume. Knowledge of these factors is necessary for the correct choice of exposure conditions and the formulation of requirements for the source material.

The answer to the question of which silicon is suitable for nuclear doping has been partially given. This answer is based on the possibility of the formation of stable oxygen complexes with radiation defects at temperatures of 400-500°C, and these donor complexes are characterized by a low level of occurrence. There is also evidence that, in addition to the indicated donors, electrically inactive oxygen clusters are formed at higher temperatures (in the range of 600–900°C), and electrically inactive SiO_2 precipitates in the region of 1000°C [46].

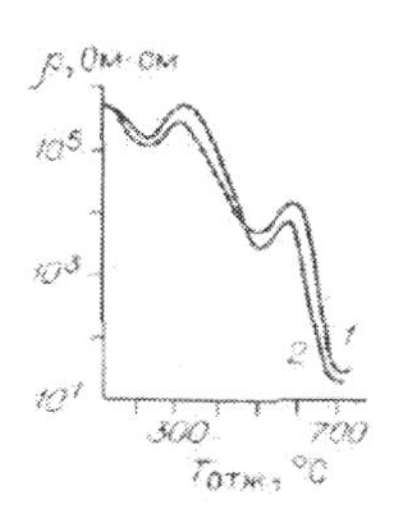

Figure 4.38. Change in resistivity during isochronous annealing of silicon irradiated by reactor neutrons with different oxygen contents (*1* - $3{,}5\cdot10^{17}$, *2* - $7\cdot10^{17}$ cm^{-2}).

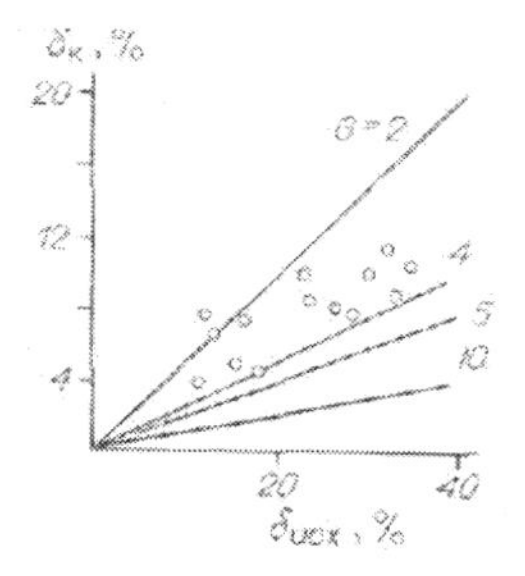

Figure 4.39. The dependence of the final variation in resistivity on the source for different ratios of input and initial concentration of phosphorus

In addition, the conductivity of the NDS [33], as well as the difference between the calculated and measured values of the concentration of carriers (see Fig. 4.38) depends on the oxygen concentration in the samples. For silicon obtained by the Czochralski method, this difference is the greater, the higher the oxygen concentration (fig. 4.27). Therefore, along with the appearance of “excess” donors, their inhomogeneous distribution is observed, due to the uneven distribution of oxygen in the initial samples. Since silicon obtained by the crucibleless melting and Czochralski methods differs primarily in the oxygen content [62], in nuclear doping, silicon grown by the crucibleless zone melting method is mainly used.

To exclude the influence of the inhomogeneous distribution of electrically active impurities in the initial silicon on the final distribution of resistivity in the NDS, it is necessary that its concentration be at least an order of magnitude lower than the introduced concentration of phosphorus. However, in practice it is not always possible to realize this condition. Therefore, the variation in resistivity in the initial silicon is limited. To quantify the admissible spread, as in [54], we use the doping degree G equal to the ratio of the final concentration of phosphorus to the residual concentration of boron or phosphorus (more precisely, their difference N_a - N_d or N_d - N_a) (see also § 2.5).

In the future, we assume that phosphorus predominates in the starting material. In this case, provided that the input concentration of phosphorus is absolutely uniform, the value of G will be equal to the ratio

$$G = N_k / N_o \tag{4.15}$$

where N_k and N_o - final and average values of the initial concentration of phosphorus, respectively. Note that in (4.15) and below, N_k and N_o can be replaced with a small error by the corresponding resistivity values.

Then it is easy to show the relationship of the scatter of resistivity along the ends or volume before and after alloying:

$$\delta_k = \delta_o / G \tag{4.16}$$

where δ_o and δ_k - variations in resistivity before and after doping, respectively.

A series of curves of the dependence of δ_k on δ_o for different values of the degree of doping G is shown in Fig. 4.39. It can be seen that in order to achieve extreme uniformity in the distribution of phosphorus concentration at sufficiently large values of the initial spread, it is necessary to increase the degree of purification of the initial material or, for a fixed value of the residual phosphorus concentration, limit the spread of the initial impurity concentration. For most practically important cases, it is sufficient to have a degree of doping equal to 5 and δ_o of the order of ±20%. The same figure shows the experimental values of δ_k for G=4.

The numerical values of δ_k after doping are found from the relation

$$\delta_k = (\rho_{\max} - \rho_{\min}) / (\rho_{\max} + \rho_{\min}) \qquad (4.17)$$

where $\rho_{\max}$ and $\rho_{\min}$ - maximum and minimum measured resistivities in the NDS ingot. It was found that in some cases the experimental values of δ_k noticeably differ from the calculated ones. This fact is explained by the fact that part of the introduced phosphorus loses its donor properties upon capture by growth defects [35]. This effect is especially pronounced when studying the microdistribution of resistivity in silicon samples, characterized by a pronounced inhomogeneous distribution of growth defects. The microphotograph of the sample on which the ρ distribution was taken with a current spreading method with a step of 20 μm is shown in Fig. 4.40. The microdistribution ρ for the same sample is shown in Fig. 4.41. The coincidence of the curve maxima in Fig. 4.41 with places of localization of defects in Fig. 4.40. Thus, along with fluctuations of the residual electrically active impurity, growth defects have a significant effect on the accuracy of resistivity falling within specified limits and the uniformity of its distribution.

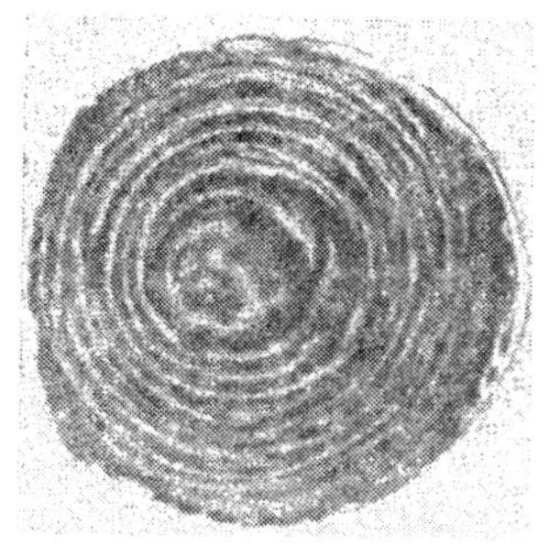

Figure 4.40. Distribution of growth defects in silicon grown in vacuum by the crucibleless zone melting method ((111) plane, sample diameter 43 mm).

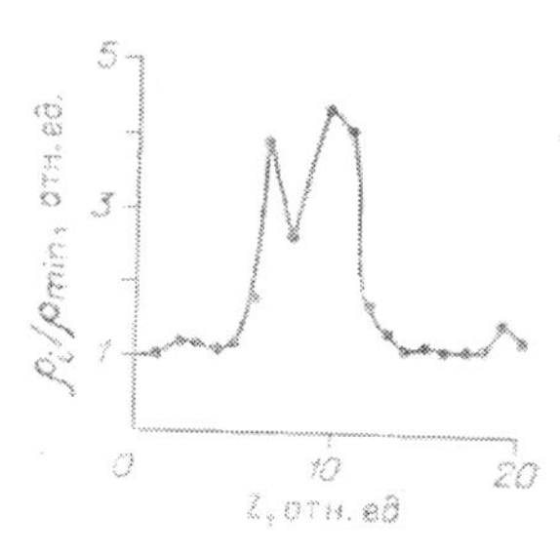

Figure 4.41. The radial microdistribution of resistivity, measured on the same sample as in fig. 4.40.

It should be noted that in addition to oxygen, phosphorus, and boron, other impurities, in particular carbon, also noticeably affect these parameters [63]. The mechanism of this effect is similar to the oxygen one — in a certain range of annealing temperatures, carbon forms complexes with radiation defects. These complexes are recombination centers and very efficiently capture electrons. Due to the fact that carbon is distributed unevenly, with incomplete annealing of these complexes, one can expect the appearance of local sites with a significantly different resistivity from the average. Defects with carbon are annealed in the region of 1000 K. This also leads to the conclusion that the choice of annealing modes should be treated very carefully.

The expected quality of nuclear doped silicon will not be achieved both at too low annealing temperatures and at obviously high ones. In the first case, the effects of irradiation will not be completely removed, in the second case, diffusion of impurities, including dopants, and their localization in the region of structural imperfections already become possible.

It was noted above that for the successful implementation of the method of nuclear transmutations it is necessary to know with absolute accuracy the absolute value of the flux density and the distribution of neutrons in the irradiation zone, as well as to accurately maintain and (or) measure the neutron flux during irradiation. In our opinion, these factors are significant, but they do not ultimately determine the quality of the NDS, because technical means are already available or at least developed to reduce their influence to a controlled limit (see § 4.1).

Let us now consider the effects associated with a stable tendency to increase the diameter of doped ingots. The motion of slow neutrons in silicon can be approximately described by a simple diffusion equation known in the kinetic theory of gases. This possibility arises due to the fact that silicon is a medium characterized by a small absorption cross section and a sufficiently large neutron scattering cross section. Therefore, the condition for the applicability of the diffusion approximation is satisfied — the smallness of the change in the neutron flux density over the mean free path.

The diffusion equation has the following form (see, for example, [64]):

$$\frac{\partial \rho}{\partial t} = D\Delta\rho - \frac{\rho}{t_{capt}} + q \qquad (4.18)$$

where $\rho_i(r, t)$ - thermal neutron density at point r at time t; Δ - Laplace operator; D - diffusion coefficient; t_{capt} - average thermal neutron lifetime; q - thermal neutron source density.

Equation (4.18) expresses the balance of the change in neutron density over time ($d\rho/dt$) under the influence of three processes: the influx of neutrons from neighboring regions ($D\Delta\rho$), absorption ($-\rho/t_{capt}$) and the formation of neutrons (q).

The solution of equation (4.18) in the case of a planar source of thermal neutrons and a semi-infinite medium is the expression

$$\varphi(x) = \varphi_0 \exp(x / L) \qquad (4.19)$$

similar to (2.27). The dependence graph (4.10) is shown in Fig. 4.42. In those cases when the radius of the sample is much smaller than the diffusion length L, the bulk effect can be neglected. As the radius of the samples increases, the decrease in the neutron flux towards the middle of the ingot can become noticeable.

To estimate the volume effect, we use the relation obtained in [21]

$$\zeta = \frac{\Phi(r)}{\Phi(0)} \simeq 1 + \frac{1}{4}\left(\frac{r}{L}\right)^2 \qquad (4.20)$$

where $\Phi(r)$ and $\Phi(0)$ - flows on the periphery and in the center of the ingot; r - is the radius of the crystal; L - is the diffusion length, which is 22.2 cm for silicon (see [65] and table 2.7). It follows from (4.20) that for r=L, ζ=1,25. The value of ζ differs by 0,05 from unity when r~0,3L. In the first case, the radius of the ingot should be about 100 mm, in the second - about 60 mm.

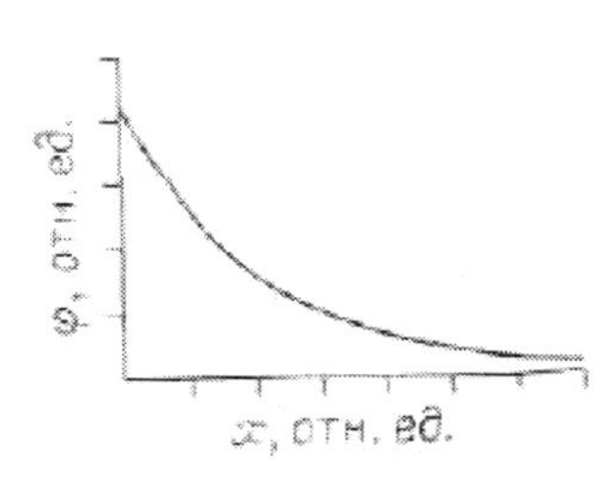

Figure 4.42. Change in the neutron flux density generated by an infinite flat source.

Thus, elementary estimates show that the volumetric effect should significantly affect the uniformity of the distribution of resistivity when doping ingots with a diameter of more than 100 mm. It should be noted that, in dealing with single crystal silicon, the indicated value must be increased at least twice (more than 200 mm), since the diffusion range L in single and polycrystals differs significantly due to different scattering cross sections of slow neutrons. More detailed studies of the diffusion mean free path of neutrons in silicon under nuclear doping are given in the next section.

4.7 Neutron flux density distribution in silicon ingots

As the size of the doped semiconductor crystals increases, the degree of uniformity of the distribution of the introduced impurity, as mentioned in Chapter 2, decreases due to self-screening of the neutron flux or gamma radiation by the sample. Therefore, in the practical implementation of the method of nuclear transmutations as applied to silicon, it is necessary to know and take into account the nature of the distribution of the flux density of slow neutrons in the irradiated samples.

Features of the neutron distribution can be determined by knowing the total interaction cross section and its components - the absorption cross section and the scattering cross section. For materials in which the scattering cross section exceeds the absorption cross section, when analyzing the distribution, it is also necessary to take into account the state of the crystal structure and the temperature of the sample at which irradiation is carried out. Next, we present the results of measurements of the total cross section for the interaction of neutrons with silicon nuclei depending on the neutron energy, temperature of the sample, and its crystal structure, performed in [66].

The intensity of the incident neutron beam passing through a cylindrical silicon sample was measured experimentally, depending on their energy. The interaction cross section was found from the relation

$$I / I_0 = \exp(-N\sigma x) \tag{4.21}$$

where I_o and I - beam intensity before and after the passage of the sample, respectively; N - number of silicon cores per unit volume; σ - interaction cross section; x - sample length.

In the studied range of neutron energy in polycrystalline samples, the change in flux density, as expected, is independent of energy. In single crystal samples, the attenuation is different for groups of neutrons of different energies. The maximum attenuation is observed for neutrons with energies of ~0.4 eV and higher. With a decrease in neutron energy, the attenuation decreases markedly.

It should be noted that the attenuation of the neutron flux in our experiment was mainly determined by scattering processes, because, firstly, for silicon, $\sigma_a < \sigma_s$ and, secondly, the geometry of the experiment is such that for a neutron one scattering event is sufficient so that it is not registered by the detector. Thus, from the obtained experimental data, one can find the value of the transport length λ_{tr} which for silicon is close to the scattering length λ_s. Specific data are given in table 4.13.

Table 4.13. Nuclear-physical characteristics of poly- and single crystal silicon

Neutron energy, eV	Monocrystalline Silicon, T = 300 K				Polycrystalline Silicon, T=300 K	
	σ_t, b	σ_a, b	σ_s, b	λ_{tr}, cm	σ_t, b	λ_{tr}, cm
0,4	1,68+0,08	0,04	1,64±0,08	11,9	2,25	11
0,0054	0,54±0,02	0,11	0,43±0,02	37,0		
0,0256	0,42±0,02	0,16	0,26±0,02	47,6		
0,016	0,40±0,02	0,20	0,20±0,02	47,7		

The nature of the change in the total cross section σ_t for poly- and single crystal silicon is shown in Fig. 4.43. For a polycrystal, σ_t is practically independent of energy and quite well coincides with the known data [65]. The situation is different for single crystal samples - for neutron groups with energies >0.4 eV, σ_t coincides with σ_t for a polycrystal; as the neutron energy decreases, the σ_t values decrease from 2 to 0.45 b (1 b = 10^{-28} m^2). The calculation results of σ_t, σ_s, σ_a for poly- and single crystal silicon are summarized in table 4.13. In the calculations, it was assumed that σ_a varies as $E^{-1/2}$ depending on the neutron energy. The absolute value of σ_a, equal to 0.16 b at E=0.025 eV, was taken from [65].

The different behavior of the dependence of σ_t on E in poly- and single crystals is explained by the fact that different scattering processes are realized in them. Taking into account coherent and incoherent elastic scattering and the fact that $\sigma_s >> \sigma_a$ in silicon, we can talk about the predominance of effects due to neutron diffraction in a polycrystal in the studied energy range. The character of the course of the dependence of σ_t on E confirms this assumption quite well.

In contrast to a polycrystal, the attenuation of the neutron flux in a single crystal occurs mainly due to inelastic scattering. The contribution of elastic coherent scattering to the overall attenuation balance is small due to the fact that in the oriented single crystals in this experiment geometry, only for certain groups of neutrons the Bragg reflection conditions are satisfied. The values of the diffusion length calculated for the polycrystal based on the table data are in good agreement with the published data (L=22 cm [65]) and significantly differ from the L value for the single crystal, which according to our data turned out to be 44 cm.

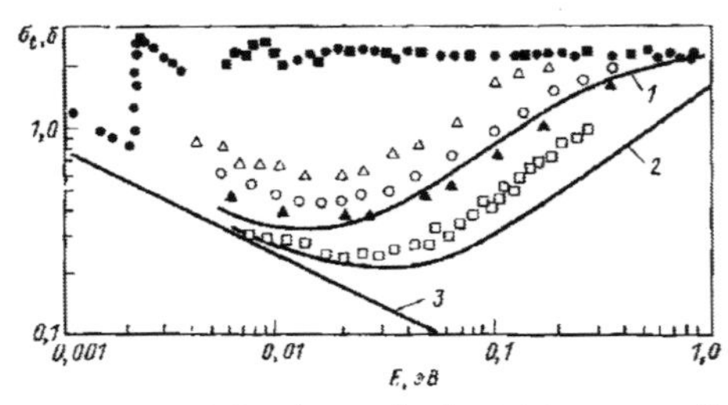

Figure 4.43. The dependence of the change in the total cross section on the energy of neutrons in poly- and single-crystal silicon samples at temperatures of 700 (Δ), 500 (O), 300 (▲), 80 K (P): ■ - experimental data for polycrystalline silicon ; · - data [65]; 1, 2 — calculation according to [67] for single-crystal samples; 3 - change in σ_a depending on the neutron energy [67]. The size of the symbols corresponds to the statistical measurement error

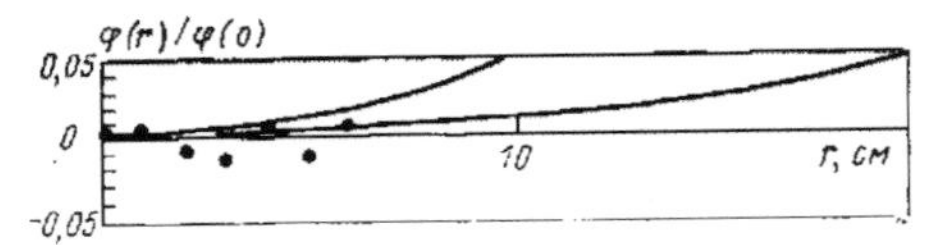

Figure 4.44. Dependence of the relative distribution of the neutron flux density on the radius of poly- (I) and single-crystal (2) silicon samples: · - experiment; - - payment.

The calculation results L were used to find the distribution of the slow neutron flux density in cylindrical samples along the z axis, which coincides in direction with the ingot forming, and along the radius of the two-zone cylindrical cell "silicon sample - water". In the first case, the calculation was carried out according to the formula [68]

$$\varphi(0,z)=\sum_{n=1}\frac{2\varphi_0}{j_n J_1(j_n)}\frac{\sinh(\gamma_n(H-z))}{\sinh(\gamma_n H)} \qquad (4.22)$$

where $\gamma_n = (j_n/R)^2 + 1/L^2$; J_1 - first-order first-order Bessel function; j_n - n root of the first order Bessel function of the first kind; φ_0 - slow neutron flux intensity at the end of the sample; H and R – extrapolated sample length and radius [68]; L - diffusion length of slow neutrons. The results of calculations of the attenuation of the flux density in poly- and single crystal samples are in good agreement with the experimentally measured attenuation of the neutron flux depending on the length of the samples.

The radial distribution of the flux density can be found in the P_1 approximation from the relation [69]

$$\frac{\varphi(r)}{\varphi(0)} = \frac{AI_0\left(\frac{r}{L}\right)\Sigma_c + \xi\Sigma_s}{A\Sigma_c + \xi\Sigma_s} \tag{4.23}$$

where A - coefficient determined from boundary conditions; Σ_c - macroscopic cross section for thermal neutron capture by sample material; $\xi\Sigma_s$ - retarding ability of sample material; I_0 - zero order Bessel function of imaginary argument.

Expression (4.23) can be greatly simplified. To do this, we expand the function $I_0(r/L)$ in a series and, since for real silicon samples $r/L<<1$, we restrict ourselves to only the first two terms. Considering also that $\xi\Sigma_s/\Sigma_c\sim 0,1$, in the final analysis, we obtain the expression [21, 70].

The character of the distribution of flux density in the radial direction in poly- and single crystal samples is shown in Fig. 4.44. It follows from this conclusion, which is very important for doping practice, that when irradiating single crystal silicon samples with a diameter of up to 200 mm with slow neutrons, attenuation of the neutron flux can be neglected $[\varphi(r)/\varphi(0)\sim 1\%]$, while attenuation the neutron flux by a polycrystalline sample becomes already noticeable and reaches 5%. It should be noted that these values may vary depending on the temperature of the irradiated samples. This is due to the fact that σ_s, which is an integral part of σ_t, depends on temperature, and L, in turn, depends on the latter, (see Fig. 4.43).

4.8. Nuclear Doped Silicon Radioactivity

As noted, in silicon it is impossible to cause the appearance of a large accumulation of phosphorus by reaction (2.31). The amount of the unstable P^{32} isotope formed depends on the neutron flux density and time as follows:

$$\left[P^{32}\right] = \left[\varphi^2\sigma_1\sigma_2\left[Si^{30}\right]\{(\lambda t - 1) + \exp(-\lambda t)\}\right]\lambda^{-2} \tag{4.24}$$

This expression is a solution to the system of equations (4.14) with respect to $[P^{32}]$.

The induced volumetric specific radioactivity of silicon, depending on the final specific resistance and flux density, is shown in Fig. 2.7. Noticeable radioactivity appears when alloying silicon samples at resistivities below 10 Ohm.sm.

According to sanitary standards [71], a safe limit for radioactive drugs, including P^{32}, is set to 10 μCi. With such total activity (or lower), it is believed that the induced activity is absent in the samples, and they can be treated as stable isotopes. In Figure 4.45 shows the amount of nuclear doped silicon that is acceptable to have at the workplace depending on the exposure time of the samples after irradiation so that the total activity of such silicon does not exceed 10 μCi. It also follows from this figure that, in principle, silicon can be doped to any resistivity value. The only difficulties are that samples with too low resistance must be kept for a very long time after irradiation (Fig. 4.46 from [72]) or they must be operated on with small amounts of silicon.

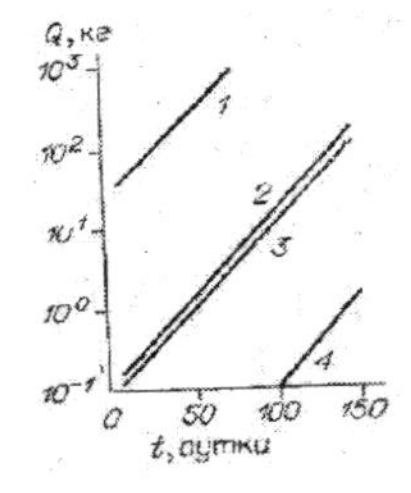

Figure 4.45. The permissible amount of silicon at the workplace doped with different resistivities, depending on the exposure time after irradiation [1]. Doping, ohm cm: 1 - from 1000 to 10; 2 - from 1000 to 1; 3 - from 10 to 1; 4 - from 1000 to 0.1

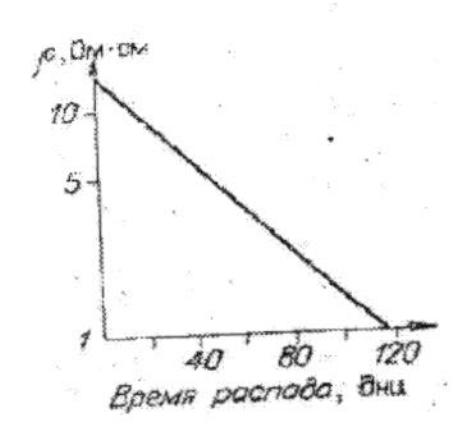

Figure 4.46 The ratio of the resistivity of silicon samples with the decay time of the induced radioactivity (P^{32}) to the maximum permissible level ($2 \cdot 10^{-4}$ μCi/g) [72].

In conclusion, we note that silicon doped with nominal values of 10 Ohm.cm and above, from the point of view of volumetric radiation caused by P^{32}, poses practically no danger, especially when you consider that the devices use discrete samples or plates of small volume.

However, a different situation arises in the analysis of surface radioactive contamination, since the indicated time is not enough for it to fall to an acceptable level due to the fact that some unstable isotopes present on the surface have a significantly longer half-life than silicon and phosphorus.

To clean the surface of the ingots from radioactive contamination, decontamination is necessary. Obviously, the choice of the cleaning method and its effectiveness depend on the nature of the forces of interaction of radioactive elements with the silicon surface and the depth of their penetration into the surface layer.

In [73], the studies were performed using single crystal silicon obtained by method the crucibleless zone melting with a specific electrical resistance of ~ 1000 Ohm·cm. Irradiation with different fluences of thermal neutrons from 10^{16} to $2 \cdot 10^{19}$ neutrons/cm^2 was carried out in the experimental channel of the WWR-t nuclear reactor, filled with water, while some samples werẹ in sealed quartz ampoules.

The impurity composition and surface contamination value of irradiated silicon samples was investigated using the activation analysis method.

In preliminary experiments, it was found that the water in the process channel is usually saturated with corrosion products of applied structural materials: stainless steel and aluminum alloy of the CAB type. Radioactive isotopes ^{24}Na, ^{51}Cg, ^{54}Mn, ^{59}Fe, ^{60}Co, ^{65}Zn, ^{95}Zr, ^{122}Sb, ^{124}Sb, ^{140}La, ^{198}Au and others are found in appreciable quantities. Almost the same elements are also present on the surface of irradiated silicon samples. Thus, it was confirmed that the most powerful source of contamination with radioactive elements of the surface of the ingots is the radiation medium in the reactor. This conclusion is also confirmed by the results of a control experiment in which samples were irradiated in evacuated quartz ampoules. The surface radioactivity of such samples was close to the background level.

After irradiation in a channel filled with water, some samples were washed in distilled water with and without ultrasound. The washing efficiency in the first case increased slightly, however, the value of surface contamination still remained noticeable, significantly exceeding the permissible norms. This fact can be explained by the fact that, firstly, chemisorption predominates during the irradiation process and, secondly, adsorbed elements can penetrate into the surface region of the irradiated sample. It was possible to identify both of these factors in the following experiment.

Several groups of samples were irradiated with fluences from 10^{16} to 10^{19} neutrons/cm^2. After irradiation, the samples were washed in an acid-peroxide solution of the composition H_2O_2 (30% solution):$HN0_3$ (60% solution):H_2O=1:1:1 at 75-85°C. This solution was chosen in due to the fact that it is very effective and does not dissolve silicon. After analysis of the impurity composition, we conducted controlled etching of the samples in the etchant $HN0_3$:HF:CH_3COOH=7:1:1 at room temperature. The thickness of the etched layer was determined by weighing the samples (~ 0.75 μm/min). To avoid secondary planting of radioactive elements from the etching solution onto the surface, the samples were washed each time in an acid-peroxide solution. Etching was carried out several times until the residual radioactivity reached the background level.

Figure 4.47 shows the activity of adsorbed radioactive elements located on the surface and in depth of silicon samples irradiated with various neutron fluences for various decontamination methods.

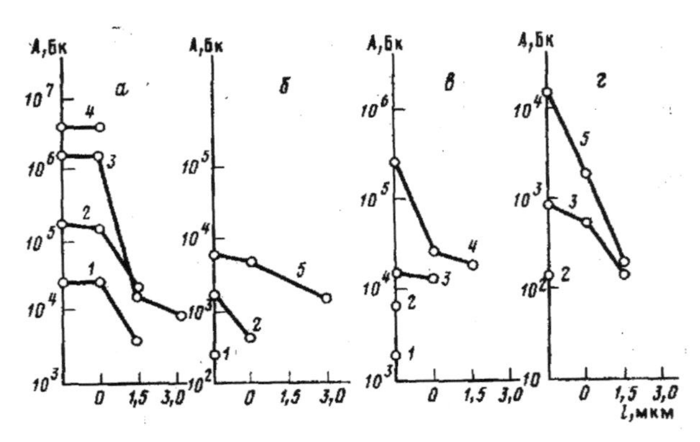

Figure 4.47 Changes in the radioactivity of the isotopes ^{24}Na (a), ^{198}Au (b), ^{140}La (c), ^{95}Zr (g) adsorbed by the silicon surface upon irradiation in the water of the WWR-t reactor channel with a flux of $2.5 \cdot 10^{13}$ neutrons/(cm^2·s) 1 (1), 10 (2), 20 (3), 50 (4) and 100 h (5) after 15 minutes of washing in an acid-peroxide solution and subsequent etching (starting from 1.5 μm)

It is seen that in the acid-peroxide solution, lanthanum and zirconium (c, d) are completely desorbed from the surface of the samples irradiated with a fluence up to $5 \cdot 10^{18}$ neutrons/cm^2. With increasing fluence, the washing efficiency decreases, since a certain fraction of adsorbed atoms penetrates into the near-surface region during irradiation. These elements are removed along with the base material by repeated bleeding.

Unlike lanthanum, zirconium, etc., rapidly diffusing elements - sodium and gold - penetrate the bulk of the material even at low neutron fluences (a, b). The different nature of the distribution of sodium and gold in depth depending on the fluence is noteworthy. After washing in an acid-peroxide solution, the residual radioactivity due to the sodium isotope changes insignificantly, which indicates a small amount of sodium on the surface of the sample. It is almost completely in volume. It can be assumed that the sodium adsorbed at the initial moment diffuses into the sample volume during the collection of fluence; subsequent sodium adsorption is apparently hindered due to the predominant adsorption of other elements.

Unlike sodium, a large fraction of adsorbed gold is on the surface of the sample and only part of it diffuses into the volume as the fluence increases.

Thus, the presented research results confirm the fact of the active deposition of radioactive elements from the water of the technological channel onto the surface of single crystal silicon samples and their subsequent diffusion into the volume upon irradiation by neutrons in the process of nuclear doping. To completely remove surface radioactive contamination, etching of a layer up to several micrometers thick is necessary.

4.9. The use of nuclear doped silicon in the manufacture of devices

Silicon intended for the production of semiconductor devices, by its properties, must satisfy the requirements that are determined primarily by the type of devices [74, 75]. As was shown, the main advantage of NDS is the uniformity of its properties. Thanks to this, he found wide application in the production of powerful devices with a large area of active *p-n* junction.

In the manufacture, in particular, of high-voltage power devices, preference is given to *n*-type single crystal silicon obtained by the crucible-free zone melting method. If silicon grown by the Czochralski method is used, changes in its properties are possible during heat treatments in the manufacturing process of devices. Typically, these changes are associated with the formation of silicon-oxygen complexes and with the release of SiO_2 particles. Therefore, for the production of the most critical and complex devices, silicon with a minimum oxygen content is used. The value of the reverse voltage that the device must withstand determines the specific resistance of silicon for this device. It is desirable that ρ be as small as possible, since with its increase the possibility and magnitude of thermal degradation of silicon parameters increase. Taking this requirement into account, we note that the method of nuclear transmutations allows one to obtain a NDS with a resistivity that differs from a given value by no more than ± 5%. Its parameters practically do not change at various stages of high-temperature heat treatment adopted in the manufacture of high-power high-voltage devices [45].

One of the most important requirements for the source material in the manufacture of devices is a high degree of uniformity of resistivity, since regions with a local change in conductivity will determine the breakdown voltage of the *p-n* junction. This situation is shown in fig. 4.48. It compares the structures of diodes made on the basis of nuclear doped and ordinary silicon. A significant difference in the breakdown voltages of diodes based on ordinary silicon is determined by the degree of difference in the average values of resistivity in individual elements (right-handed structures). The structural elements based on the NDS (left) are characterized by higher breakdown voltages and, accordingly, completely identical parameters [76]

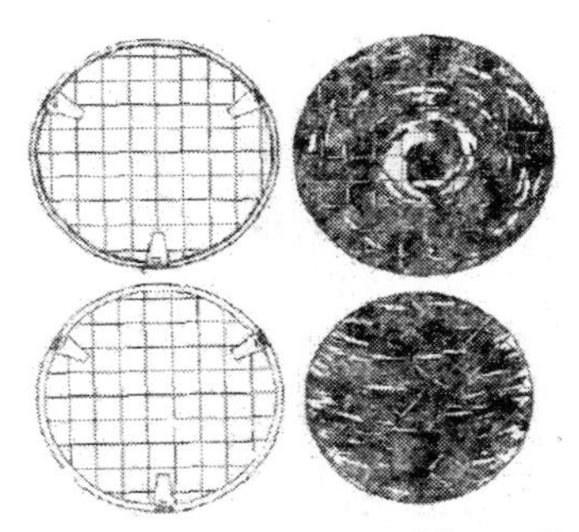

Figure 4.48, Influence of the quality of the initial silicon on the characteristics of the structures of p - n junctions [76] (explanation in the text).

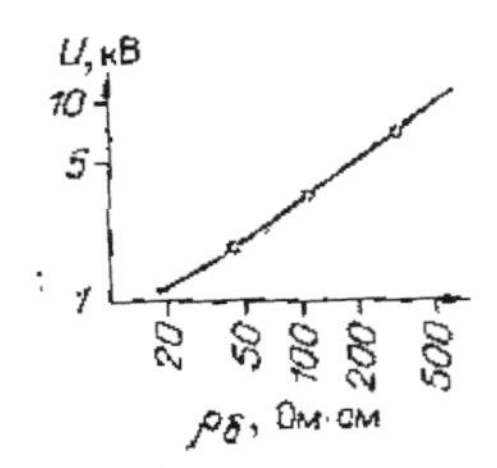

Figure 4.49. Dependence of the breakdown voltage of the collector junction on the resistivity of the base of p-n-p structures made on the basis of NDS [40]. A point is an experiment, a line is a calculation.

It should be noted that the criterion for a high degree of uniformity of the base material and its thermal stability can be considered good agreement with the calculated values of the breakdown voltage of the pn junctions made on the basis of the NDS (Fig. 4.49) [40]. The reverse branches of the current

– voltage characteristics of high voltage transistors made of ordinary and nuclear doped silicon are compared in Fig. 4.50. In the latter case, the characterization is more stringent, which indicates the absence of structural disturbances and local variations in resistivity, in the region of which electric field inhomogeneities arise [77].

In this case, the parameters of devices made of NDS are more thermostable than similar parameters of devices made of ordinary material (Fig. 4.51). The latter once again demonstrates the significant advantages of NDS over ordinary [40, 78].

Thus, the most promising material for the production of power devices with a large active area that can withstand large reverse voltages is silicon doped using the nuclear transmutation method.

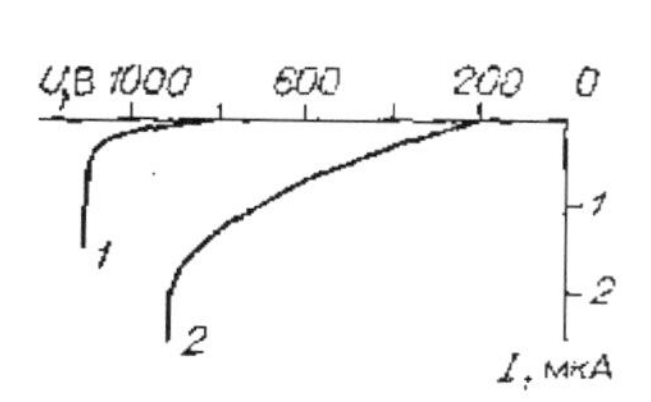

Figure 4.50. Reverse branches of the current-voltage characteristics of transistors made of nuclear doped (1) and ordinary (2) silicon.

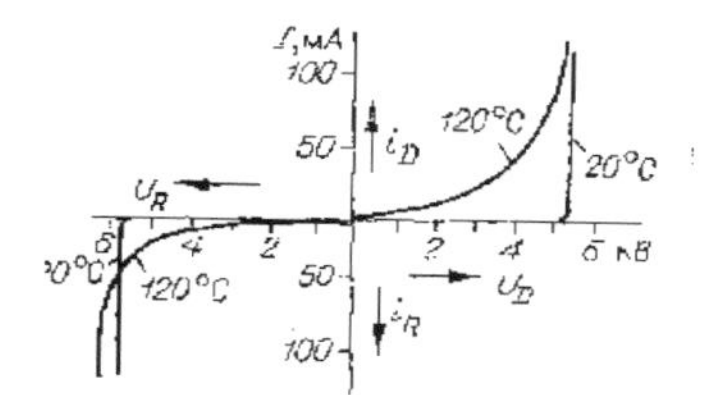

Figure 4.51. Shutoff characteristics of a 4.5 kV thyristor made of nuclear doped silicon at room temperature and at 120°C [78].

The use of NDS in the production of microelectronic devices, large and ultra-large integrated circuits, where strict requirements for the uniformity of the material on the entire plate and the slightest local fluctuations are unlikely, will be even more promising. Unfortunately, the use of NDS in microelectronics is only just beginning, and so far there is not enough experience for generalizations.

4.10. Other applications of the nuclear transmutation method

Slow neutron flux density measurement

It is known that each donor atom in a semiconductor supplies one electron to the conduction band. The concentration of current carriers can be determined with a sufficient degree of accuracy from the relation

$$R_x = \pm \frac{r}{ne} \quad (4.25)$$

where R_x - Hall constant; n - carrier concentration; e - electron charge; ± used depending on media type; r - constant. From experience determine the Hall constant [41]. Taking into account (4.25) and (4.2), we find the flux density of slow neutrons from relation (2.24). According to [79], the error in determining the flux density mainly depends on the error in measuring the value of R_x and is no more than 10%.

The main complicating factor in determining the flux by this method is radiation defects in semiconductors, which should be annealed before measuring the Hall constant (see Chapter 3).

The results of measurements of the slow neutron flux density in one of the experimental channels of the WWR-t reactor are shown in Table 4.14. For comparison, the data obtained by the activation method of the foils are also indicated. From table 4.14 it follows that the values of the slow neutron flux calculated on the basis of Hall effect phosphorus concentration measurements for six different silicon samples and the known irradiation time are in good agreement with each other, as well as with the flux value obtained by the foil activation method.

Table 4.14. The results of measurements of the slow neutron flux in the experimental channel of the WWR-t reactor according to the Hall effect in silicon and according to the foil activation method

sample number	before exposure		preset integral neutron flux, cm^{-2}	after exposure		neutron flux φ, $cm^{-2} \cdot s^{-1}$	
	type of current carriers	carrier concentration, cm^{-2}		type of current carriers	carrier concentration, cm^{-2}	Hall effect	foil method
1	p	$3{,}5\cdot 10^{12}$	$5{,}0\cdot 10^{17}$	n	$8{,}0\cdot 10^{13}$	$6{,}2\cdot 10^{13}$	$6{,}2\cdot 10^{13}$
2	p	$2{,}6\cdot 10^{12}$	$5{,}0\cdot 10^{17}$	n	$8{,}0\cdot 10^{13}$	$6{,}2\cdot 10^{13}$	$6{,}2\cdot 10^{13}$
3	p	$1{,}2\cdot 10^{12}$	$1{,}3\cdot 10^{18}$	n	$2{,}1\cdot 10^{14}$	$6{,}1\cdot 10^{13}$	$6{,}2\cdot 10^{13}$
4	p	$1{,}3\cdot 10^{12}$	$1{,}3\cdot 10^{18}$	n	$2{,}1\cdot 10^{14}$	$6{,}1\cdot 10^{13}$	$6{,}2\cdot 10^{13}$
5	p	$1{,}2\cdot 10^{12}$	$2{,}0\cdot 10^{19}$	n	$1{,}4\cdot 10^{15}$	$6{,}2\cdot 10^{13}$	$6{,}2\cdot 10^{13}$
6	p	$1{,}2\cdot 10^{12}$	$2{,}0\cdot 10^{19}$	n	$1{,}4\cdot 10^{15}$	$6{,}2\cdot 10^{13}$	$6{,}2\cdot 10^{13}$

Currently, industry produces a number of semiconductor materials into which, like silicon, using the method of nuclear transmutations, donor or acceptor impurities can be successfully introduced. Therefore, it is possible to select the appropriate semiconductor for dosimetry over a wide range of the neutron flux, as well as the flow of charged particles and bremsstrahlung γ radiation. One of the advantages of the described method of measuring particle flux is that the dosimetric information in the samples is stored indefinitely. In some cases, this factor is of fundamental importance, for example, when in another way it is necessary to obtain the necessary information when measuring the induced activity of short-lived isotopes, and also when it is necessary to reproduce this information again at any time after irradiation.

Creating resistivity reference samples

Currently, a very wide range of semiconductor materials of a wide variety of types is being produced. The circle of consumers of semiconductor materials is constantly expanding, international integration and cooperation in the field of semiconductor instrument making and the production of source material is increasing. In this regard, the introduction of common tools and methods for measuring the electrophysical parameters of semiconductor materials is of great importance. So, to measure the electrical resistivity of single crystal silicon and evaluate its quality by this parameter, the four-probe method is mainly used [80]. For regular verification of measurement accuracy and certification of four-probe installations, standard samples are required.

Nuclear doped silicon was proposed to be used as such samples in [81]. For nuclear doping, *p*-type silicon samples with a specific resistance of 5000 Ohm·cm were selected. The neutron flux density in the working zone of the reactor was 10^{12} $cm^{-2} \cdot s^{-1}$. Samples were irradiated 22; 110; 220 and 440 hours. After irradiation, they were subjected to high-temperature annealing, as a result of which the resistivity of the samples was (in decreasing order of irradiation time) 1,4; 2,8; 5,6 and 27 Ohm·cm. To assess the uniformity, the resistivity was measured by four- and three-probe methods. It was established that the standard deviation of the resistivity, measured by the four-probe method, from the average value is comparable with the instrument error of the equipment used. In this regard, NDS is the most promising material for the production of standards of resistivity, which can be used for calibration and non-contact methods for measuring resistivity. A description of the standards and their nomenclature can be found in [82].

Creating *p-n* transitions

The principle of creating *p-n-* junctions using the method of nuclear transmutations, as shown in Chapter 2, consists in creating such irradiation conditions under which slow neutrons fall only in predetermined regions of the sample. To do this, use special screens made of materials that effectively absorb thermal neutrons. As such screens, a natural mixture of cadmium and gadolinium isotopes is predominantly used, as well as cadmium-113, boron-10 isotopes. Natural cadmium is used in the production of *p-n* junctions, *p-n-p-* and other simple structures. In the manufacture of microcircuits, cadmium-113 or gadolinium is preferred as protection.

In general terms, radiation technology for the manufacture of semiconductor devices includes the following operations [83-86]: development and manufacture of screens of the necessary configuration; capsule design and manufacture; preliminary measurements and selection of exposure conditions; capsule irradiation and neutron flux monitoring; post-radiation processing and measurements; manufacturing of devices, including metallization, oxidation, creation of additional contacts.

The simplest scheme for obtaining the *p-n* junction is shown in Fig. 4.52, and in fig. Figure 4.53 shows the assembly with which *p-n* junctions of a simple configuration were obtained (Fig. 4.54).

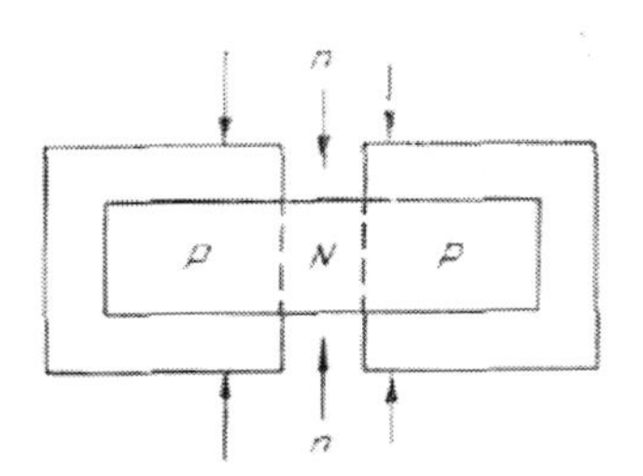

Figure 4.52. The scheme for obtaining *p-n* junctions

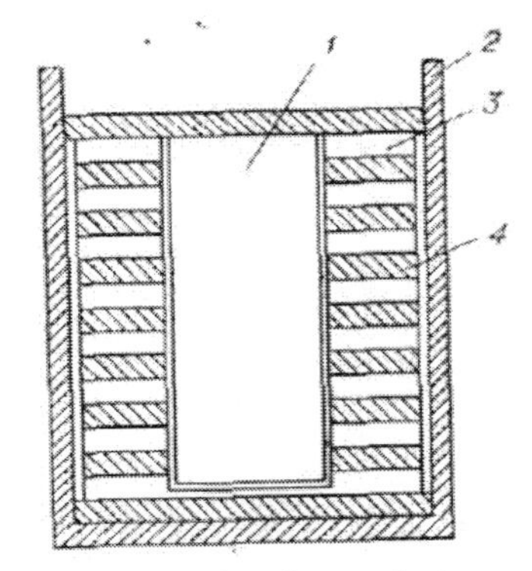

Figure 4.53. Device for producing *p-n* junctions. 1- silicon sample; 2 - capsule; 3 - aluminum gasket; 4 - cadmium screen

The main material used for the manufacture of assemblies is aluminum or alloys based on it. Assemblies are designed so that they can be easily disassembled and assembled using remote-controlled mechanisms.

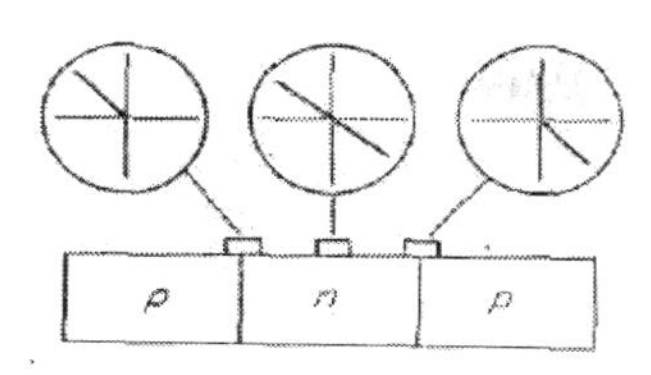

Figure 4.54. Current-voltage characteristics of samples with *p-n* junctions.

The choice of exposure conditions is reduced to the fulfillment of conditions (2.47), (2.48) and (2.55). With the known hole conductivity in the starting material and the given ratio between the carrier concentrations in the *n*- and *p*-regions of the transition, by choosing the appropriate protection material, the gap geometry, and the neutron spectrum, the required carrier concentrations in the corresponding regions can be obtained. As indicated in Chapter 2, to create a sharp *p-n* junction, a very soft neutron spectrum is required, which in the general case can be obtained, for example, using a heat column, however, the absolute value of the thermal neutron flux density becomes so small that it takes an irradiation time to reach the required phosphorus concentration going beyond reason. Therefore, radiation is usually carried out by a neutron flux with a cadmium ratio of the order of 30. According to [84], the resistance ratio that can be achieved lies in the range from 1: 1 to 50: 1. The diodes made in this way had a breakdown voltage of 25 to 50 V.

The transition area reached 50 mm^2. After irradiation, the samples were decontaminated and annealed at a temperature of 800°C.

Using the method of nuclear transmutations, in addition to diodes, transistors and various integrated circuits are manufactured. If you create an oxide of controlled thickness on samples with an *n-p-n* structure, then, in principle, you can get a transistor, the circuit of which is shown in Fig. 4.55. The central region serves as a gate, part of one, freed from oxide, the n-region acts as a sink, and part of the other acts as a source.

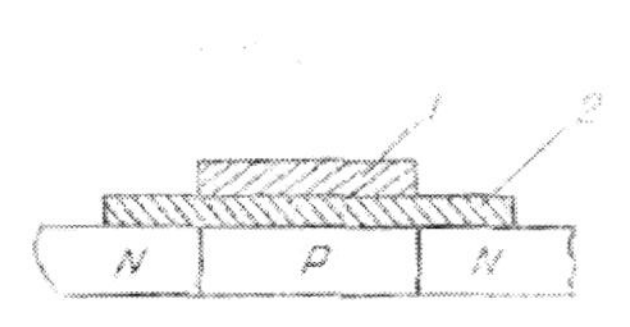

Figure 4.55. Scheme of a transistor structure with active oxide.
1 - metal; 2 - oxide.

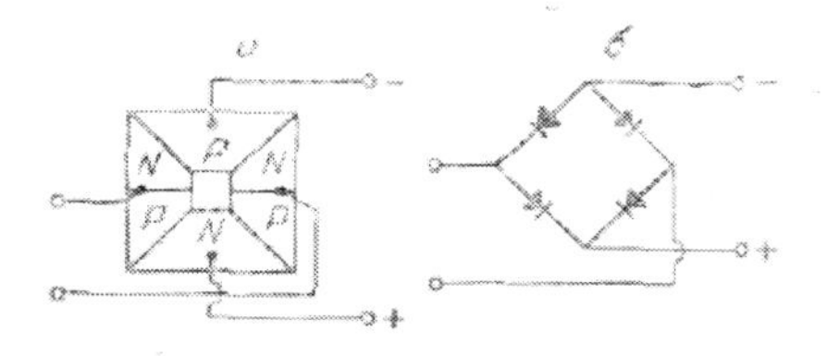

Figure 4.56. The structure of the microcircuit (a) and its electrical equivalent circuit (b) [84].

Electrical measurements have shown that an "inversion" layer arises between these regions. The layer thickness is about 1 μm. A more detailed description of this structure can be found in [84]. One of the chip options is shown in Fig. 4.56. A feature of such schemes is a peculiar arrangement of *n*- and *p*-type regions. The transition itself is perpendicular to the surface of the plates.

Based on such areas, various microcircuits of the original structure were proposed in [84]. Noteworthy are also diodes fabricated by the method of nuclear transmutations, the capacitance of which sharply depends on the applied reverse voltage. Such diodes are characterized by the following parameters: capacitance at zero bias of 60 pF; breakdown voltage of 50 V; the scope of the change in capacity is 400%.

In the manufacture of high-power high-voltage devices, special measures are taken to prevent surface breakdown and migration of ionized impurities on the surface. So, guard rings are created in planar transistors, and chamfers with a positive or negative bevel angle are created in other devices [74].

As a result, the field on the surface decreases and becomes significantly lower than in volume. A similar situation can be achieved using the method of nuclear transmutations. Using special screens during irradiation, in silicon wafers, you can create any desired profile for the distribution of resistivity in diameter and thus control the depth and width of the pn junction at the periphery and in the center of the device. One of the profile options in silicon wafers of a given thickness is shown in Fig. 4.57.

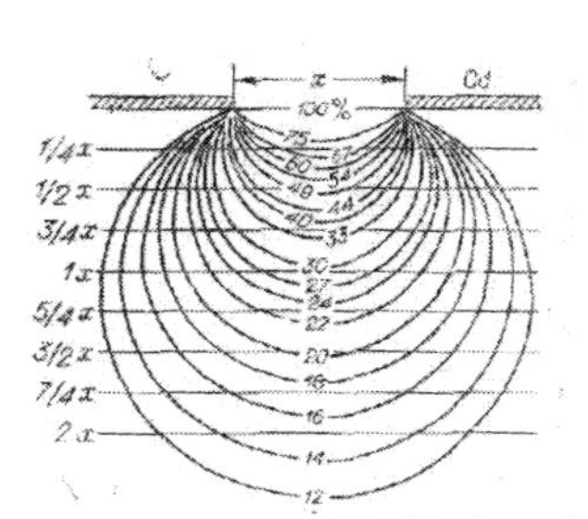

Figure 4.57. The distribution of the neutron flux at different distances from the gap.

Some trends in the development of the nuclear doping method of semiconductors

Taking into account the immediate development trends of semiconductor instrument engineering, it is possible to reliably associate the further spread and improvement of the method of nuclear transmutations with the production of primarily semiconductor silicon. So, already today there are a number of ideas regarding the expansion of the range of impurities that can be introduced into silicon using nuclear reactions.

For example, in the near future it will become realistic to obtain p-type silicon using photonuclear reactions (see section 2.3), since if there are radiation sources with the required intensity and energy of γ-quantum, it will be possible to take advantage of the experience in solving all other technological issues doping of silicon with the help of nuclear reactions on slow neutrons.

Considering also the high level of engineering and technology achieved, we can talk about doping silicon with gallium as a near term. For this, silicon, into which a certain amount of germanium is preliminarily introduced, must be irradiated with slow neutrons [87]. As follows from table 2.6, germanium undergoes nuclear transmutations with the formation of gallium by reaction

$$\mathrm{Ge}^{70}(n,\gamma)\mathrm{Ge}^{71} \rightarrow \mathrm{Ga}^{71} \tag{4.26}$$

In this case, Ga^{71} is formed simultaneously with phosphorus obtained from Si^{30}; therefore, it is necessary to create conditions under which the concentration of Ga^{71} will be higher than the concentration of P^{31} and other possible donor impurities. To increase the efficiency of gallium formation, silicon can be doped with germanium preliminarily enriched in the Ge^{70} isotope, or germanium can be introduced into silicon without the Si^{30} isotope. Similarly, n- or p-type germanium can be obtained by nuclear doping a material enriched in one of the isotopes Ge^{70}, Ge^{74}, Ge^{76}, and introducing Ga^{71}, As^{70}, and Se^{77} impurities, respectively, without the complicating effect of competing nuclear reactions.

The use of isotopic compositions is also of interest from the point of view of elucidating the contribution of nuclear transmutations to defect formation and other processes associated with the formation of the properties of nuclear doped materials. Therefore, it seems tempting to conduct a comparative comprehensive study of silicon samples irradiated under identical conditions, which are: 1) a natural mixture of Si^{28}, Si^{29}, Si^{30} isotopes; 2) a mixture of isotopes Si^{28} and Si^{29}; 3) isotope Si^{30}. In general, it should be expected that the transition to nuclear doping of certain isotopic compositions can play a large role in solving the problem of nuclear doping of semiconductor and other materials and gain new knowledge about the properties of semiconductors of monoisotopic composition.

The ideas of using the nuclear transmutation method for doping epitaxial layers seem very promising [88]. In this case, it is possible to circumvent the limitations of the method due to the small mean free path of the irradiating particles in the semiconductor and use it to obtain semiconductor layers, in particular, from the $A^{III}B^{V}$ family. The use of masks will allow controlled doping of individual areas of the epitaxial film.

In the practice of doping some semiconductors, the method proposed in Ref. [89] may also be useful. If for some reason the desired impurities cannot be introduced into the crystal in the usual way, then they are preliminarily introduced into the surface layer of the material by nuclear reactions. Then, during the subsequent melting of the material, the volume of the formed single crystal can be saturated with an impurity transferred from the surface layer. In some cases, by ion implantation or by epitaxial growth, atoms of such elements can be introduced into the surface layer that, when irradiated, are capable of undergoing nuclear transmutations with the formation of the necessary impurity.

In conclusion, we note that the problems solved in semiconductor instrument making are very diverse, and the methods used to solve them are accordingly diverse. The method of nuclear transmutations has already taken a worthy place among them, and one can hope both for further progress of the method itself and for an increase in its value in connection with the constant tightening of the requirements for the properties of the starting material, as well as for the parameters of devices manufactured on the basis of various semiconductor materials.

References:

1. Haas W., Schnuller M. Silicon Doping by Nuclear Transmutation. - J. Electr. Mat., 1976. Vol. 5. Issue 1. p. 57-68.
2. Neutron Transmutation Doping in Semiconductors. Proc. 2-th Intern. Conf. Columbia, Missouri, April 23-26, 1978. Ed. J. M. Meese. N. Y.- London, Plenum Press. 1979. 371 p.
3. Bat' G.A., Kochenov A.S, Kabanov L.P. *Issledovatel'skie iadernye reaktory* [Research nuclear reactors]. Moscow. Atomizdat. 1972. 272

p. (in Russian)
4. Bourdon J.L., Restelli G. An Automated Irradiation Facility for Neutron Doping of Large Silicon Ingots. – In book: Neutron Transmutation Doping in Semiconductors. Proc. 2-th Intern. Conf. Colombia, Missouri. April 23-26, 1978. Ed. J. M. Meese. N. Y. - London, Plenum Press. 1979. p. 182 - 195.
5. Gunn S.L., Meese J.M, Alger D.M. High Precision Irradiation Techniques for NTD Silicon at the University of Missouri Research Reactor. – In book: Neutron Transmutation Doping in Semiconductors. Proc. 2-th Intern. Conf. Columbia, Missouri. April 23-26, 1978. Ed. J. M. Meese. N. Y. - London, Plenum Press, 1979, p. 157-163.
6. Gres'kov I.M., Solov'ev P.P., Kharchenko V.A. *Iadernoe legirovanie poluprovodnikov* [Nuclear Doping of Semiconductors]. Series: *Radiatsionnaia stoikost' organicheskikh materialov* [Radiation resistance of organic materials]. Moscow. NIITEKhIM. 1982. p.32. (in Russian)
7. Krull W., Martens P. An Installation for the Irradiation of Large Silicon Monocsrystals for the Production of Power Thyristors - Kerntechnik, 1977. Issue 3. p. 131-135.
8. Berliner R., Wood S.A. Computer Controlled Irradiation System for the University of Missouri Research Reactor. - In: Neutron Transmutation Doping in Semiconductors. Proc. 2-th Intern. Conf. Columbia, Missouri, April 23-26, 1978. Ed. J. M. Meese. N. Y. - London, Plenum Press, 1979, p. 215-228.
9. Kharchenko V.A., Solov'ev P.P. *Radiatsionnoe legirovanie kremniia* [Radiation Doping of Silicon]. Proceedings of the USSR Academy of Sciences. Inorganic materials. 1971. Vol. 7. Issue 12. p. 2137-2141. (in Russian)
10. Galata A.Ia., Stuk A.A., Kharchenko V.A. *Optimizatsiia s ispol'zovaniem EVM protsessa oblucheniia pri iadernom legirovanii kremniia* [Computer optimization of the irradiation process during nuclear doping of silicon]. Tsvetnye metally. 1983, Issue 4. (in Russian)
11. Rvachev V.G. *Geometricheskie prilozheniia algebry logiki* [Geometric Applications of Logic Algebra]. Kiev. Tekhnika. 1967.

p.187. (in Russian)
12.Galata A.Ia., Stoian Iu.G. Kibernetika [Cybernetics]. 1972. Issue 2. p. 81-86
13.Smith G.G. Neutron Doping of Silicon in the Harwell Research Reactors. - In: Neutron Transmutation Doping in Semiconductors. Proc. 2-th Intern. Conf. Colombia, Missouri, April 23-26, 1978. Ed .J. M. Meese. N. Y. - London, Plenum Press, 1979, p. 157-163.
14.Bickford N.A., Fleming R.F. Silicon Irradiation Facilities at the NBS Reactor. - In: Neutron Transmutation Doping in Semiconductors. Proc. 2-th Intern. Conf. Columbia, Missouri, April 23-26, 1978. Ed. J. M. Meese. N. Y. - London, Plenum Press, 1979, p. 165-170.
15.Morrissey J.E., Tillinghast T. General Electric Test Reactor NTD Silicon Development Program. - In: Neutron Transmutation Doping in Semiconductors. Proc. 2-th Intern- Conf. Columbia, Missouri, April 23-26, 1978. Ed. J. M. Meese. N.Y. - London, Plenum Press. 1979. p. 171-179.
16.Abagian L.P., Bazaziants O.N., Nikolaev M.N., Tsibulia A.M. *Gruppovye konstanty dlia rascheta iadernykh reaktorov. Spravochnik* [Group constants for calculating nuclear reactors. Directory]. Moscow. Energoizdat. 1981. (in Russian)
17.Herzer H. Neutron-Doped Silicon – A. Market Review. In: Neutron Transmutation Doping Silicon, Proc. 3th Intern. Conf. Copenhagen, August 27-29, 1980. Ed. J. Guldberg, N.-Y., - London, Plenum Press, 1981. p. 1- 17
18.Jang R., Cleland W., Wood R., Abraham M. Radiation Damage in Neutron Transmutation Doped Silicon: Electrical Property Studies. - J. Appl. Phys., 1978. Vol. 49. Issue 9.
19.Messier J. Variations de resistivite Produites Dans du Silicium P de Hante Resistivite Par des Neutrons Thermiques. - P. r. Acad. Sci., 1962. Vol. 255. Issue 17. p. 2083-2085.
20.Sommer K., Sontag A. Application of New Technogics to H.V.D.C Thyristors Produstion. – In book: Vsemirnyi elektrotekhnicheskii congress. 1977. July. Moscow. Section 5A. Issue 47. Moscow. 1977.
21.Janus H., Malmros O. Application of Thermal Neutron Irradiation for Large Scale Production of Homogeneous Phosphorus Doping of Floatzone Silicon, - IEEE Trans, Electr. Dev., 1976, Vol. ED-23, Issue

8, p. 797-802.
22.Kharchenko V.A., Solov'ev P.P. *Radiatsionnoe legirovanie kremniia* [Radiation Doping of Silicon]. Semiconductor Physics and Technology. 1971. Vol. 5. Issue 8. p. 1641-1643. (in Russian)
23.Stein H. Atomic Displacement Effects in Neutron Transmutation Doping. - In: Neutron Transmutation Doping in Semiconductors. Proc. 2-th Intern. Conf. Colombia, Missouri, April 2-26, 1978. Ed. J. M. Meese. N. Y. - London, Plenum Press, 1979. p. 229-247.
24.Larson B.C, Young R.T., Naragan T. Defects Annealing Studies in Neutron Transmutation Doped Silicon. - In: Neutron Transmutation Doping in Semiconductors. Proc. 2-th Intern. Conf. Columbia, Missouri, April 23-26, 1978. Ed. J. M. Meese. N. Y. - London, Plenum Press. 1979. p. 281-289.
25.Young M.H., March O.I., Baron R. Shallow Defect Levels in Neutron Irradiated Extrinsic p-Type Silicon. - In: Neutron, Transmutation Doping in Semiconductors. Proc. 2-th Intern. Conf. Columbia, Missouri, April 23-26, 1978. Ed. J. M. Meese. N. Y. – London. Plenum Press. 1979. p. 335-343.
26.Senes A., Sifze G. Stabilization of Transmutation Doped Silicon.-In: Semiconductor Silicon, 1977. Eds. H. Huff, E. Sirtl. Princeton, The Electrochem. Soc, p. 135-141.
27.Bunting B., Verghese K., Saxe B. Effects of Thermal Neutron Irradiation in p-Type Silicon. - J. Appl. Phys. 1968. Vol. 39. Issue 1. p. 342-343.
28.Meier Dzh., Erikson L., Devis Dzh. *Ionnoe legirovanie poluprovodnikov* [Ion doping of semiconductors]. Moscow. World. 1973. 296 p. (in Russian)
29.Koval' Iu.P., Mordkovich V.II., Temper E.M., Kharchenko V.A. *Opticheskie svoistva kremniia, obluchennogo neitronami* [Optical properties of neutron-irradiated silicon]. Semiconductor Physics and Technology. 1972. Vol. 6. Iss. 7. p. 1317-1322. (in Russian)
30.Glairon P., Meese J. Isochronal Annealing of Resistivity in Floatzoneand Czochralski NTD Silicon. - In: Neutron Transmutation Doping in Semiconductors. Proe. 2-th Intern. Conf. Columbia, Missouri. April 23-26, 1978. Ed. J. M. Meese. N. Y. - London, Plenum Press. 1979. p. 219-305.

31.Gres'kov I.M., Efimovich O.N., Solov'ev P.P., Kharchenko V.A., Shapiro V.G. *Otzhig radiatsionnykh defektov v kremnii, obluchennom nei-tronami i bystrymi elektronami* [Annealing of radiation defects in silicon irradiated by neutrons and fast electrons]. Physics and chemistry of materials processing. 1976. Issue 5. p. 31-34. (in Russian)

32.Kharchenko V.A., Voronov I.N., Mordkovich V.N., Smirnov B.V., Solov'ev P.P. *Elektrofizicheskie svoistva radiatsionno-legirovannogo kremniia, poluchennogo metodom Chokhral'skogo* [Electrophysical properties of radiation-doped silicon obtained by the Czochralski method]. In book: *Vtoraia konf. po fiziko-khimicheskim osnovam legirovaniia poluprovodnikovykh materialov* [The second conf. on physicochemical principles of doping of semiconductor materials] (abstracts). Moscow 1972. (in Russian)

33.Mordkovich V.N., Solov'ev P.P., Temper E.M., Kharchenko V.A. *Provodimost' kremniia, podvergnutogo neitronnomu oblucheniiu i otzhigu* [Conductivity of silicon subjected to neutron irradiation and annealing]. Semiconductor Physics and Technology. 1974, Vol. 8, Iss. 1, p. 210-213. (in Russian)

34.Efimovich O.N., Solov'ev P.P., Kharchenko V.A. *Neitronografichsskoe issledovanie defektnoi struktury obluchennykh monokristallov kremniia* [Neutron diffraction study of the defective structure of irradiated silicon single crystals]. In book: Third All-Union Conference. Structural defects in semiconductors (abstracts). Patr. II. Novosibirsk. 1978. p. 182. (in Russian)

35.Gres'kov I.M., Solov'ev P.P., Kharchenko V.A. *Vliianie rostovykh defektov na izmenenie provodimosti kremniia, obluchennogo neitronami* [The influence of growth defects on the change in the conductivity of neutron-irradiated silicon]. Semiconductor Physics and Technology. 1977. Vol. 11. Iss. 8. p. 1598-1601. (in Russian)

36.De Kock A. Microdefects in Dislocation-Free Silicon Crystals, - Phys.Bes. Rep., 1973, Suppl. 1, p. 1—106.

37.Gres'kov I.M., Solov'ev P.P., Kharchenko V.A. *Zavisimost' nekotorykh izmenenii elektrofizicheskikh svoistv i mikrostruktury kremniia ot uslovii oblucheniia i svoistv iskhodnogo materiala* [Dependence of some changes in the electrophysical properties and

silicon microstructure on irradiation conditions and the properties of the starting material]. In book: *Radiatsionnye effekty v kremnii* [Radiation effects in silicon]. Preprint KIIaI-76-23. Kiev. 1976. p. 26-27. (in Russian)
38.Gusev V.M., Titov V.V. *Issledovanie otzhiga defektov, vvedennykh kremnii putem oblucheniia ionami Si+* [Investigation of annealing of defects introduced by silicon by irradiation with Si + ions]. In book: *Radiatsionnaia fizika kristallov i p-n-perekhodov* [Radiation physics of crystals and p - n junctions]. Minsk. Science and technology. 1972. p. 95. (in Russian)
39.Karmanov V.T., Pavlov P.V., Zorin E.I., Tetel'baum D.I., Khokhlov A.F. *Otsutstvie donornykh svoistv fosfora pri diffuzii v kremnii, podvergnutyi ionnoi bombardirovke* [Lack of donor properties of phosphorus upon diffusion into silicon subjected to ion bombardment]. Semiconductor Physics and Technology. 1975. Vol. 9. Iss. 9. p. 1780-1789. (in Russian)
40.Hill M., Van Iseqhem P., Zimmerman W. Preparation and Application of Neutron TransmulaLion Doped Silicon for Power Device Research. - IEEE Trans. Electr. Dev. 1976. Vol. ED-23. Issue 8. p. 809-813.
41.*Poluprovodniki* [Semiconductors]. Edited by N. B. Khenneia. Moscow. 1962. 667 p. (in Russian)
42.Hart R., Albert L., Skinner N. Measurement of ^{31}P Concentrations Produced by Neutron Transmutation Doping of Silicon. In book: Neutron Transmutation Doping in Semiconductors. Proc. 2-th Intern. Conf. Columbia, Missouri. April 23-26. 1978. Ed. J. M. Meese. N. Y. - London, Plenum Press, 1979, p. 345-354.
43.Smit R. *Poluprovodniki* [Semiconductors]. Moscow. 1962. 467 p. (in Russian)
44.Dirnli Dzh., Nortrop D. *Poluprovodnikovye schetchiki iadernykh izluchenii* [Semiconductor Nuclear Radiation Counters]. Moscow. Mir. 1966. 359 p. (in Russian)
45.Kharchenko V.A., Smirnov B.V., Solov'ev P.P., Fetisova G.A., Voronov I.P., Bane V.E. *Vliianie termoobrabotki na elektrofizicheskie svoistva radiatsionno-legirovannogo kremniia* [The influence of heat treatment on the electrophysical properties of radiation-doped silicon]. Proceedings of the USSR Academy of Sciences. Inorganic materials.

1971. Vol. 7. Issue 12. p. 2142-2145. (in Russian)

46. Kireev P.P. *Fizika poluprovodnikov* [Semiconductor Physics]. Moscow. Higher school. 1975. 584 p. (in Russian)

47. Mordkovich V.N., Solov'ev P.P., Temper E.M., Kharchenko V.A. *Provodimost' kremniia, podvergnutogo neitronnomu oblucheniiu i otzhigu* [Conductivity of neutron-irradiated and annealed silicon]. In book: *Radiatsionnye defekty v poluprovodnikakh (rasshirennye tezisy)* [Radiation defects in semiconductors (extended abstracts)]. Minsk. 1972. p. 92-93. (in Russian)

48. Cleland J., Fleming P., Westbrook R., Wood R, Young R. Electrical Property Studies of Neutron Transmutation Doped Silicon. In book: Neutron Transmutation Doping in Semiconductors. Proc. 2-th Intern. Conf. Columbia. Missouri. April 23-26. 1978. Ed. J. M. Meese. N. Y. – London. Plenum Press. 1979. p. 261-279.

49. Lymar' G.F. *Izmerenie udel'nogo soprotivleniia kremnievykh epitaksial'nykh sloev metodom soprotivleniia rastekaniia tochechnogo zonda* [Measurement of resistivity of silicon epitaxial layers by the method of resistance of spreading of a point probe]. - Electronic equipment. Series 2. Semiconductors. 3970. Iss. 4 (54). p. 3-5. (in Russian)

50. Afonin L.P., Mordkovich V.P., Smirnov B.V., Solov'ez P.P., Temper E.M., Kharchenko V.A. *Radiatsionno-legirovannyi kremnii dlia vysokovol'tnykh priborov* [Radiation-doped silicon for high-voltage devices]. Electronic industry. 1976. Iss. 6 (54). p. 53 - 54. (in Russian)

51. Herrmann H., Herzer H. Doping of Silicon by Neutron Irradiation. J. Electrochem. Soc. 1975. Vol. 122. Issue 11. p. 1568-1569.

52. Malmros O. The Minority Carrier Lifetime of Neutron Doped Silicon. In book: Neutron Transmutation Doping in Semiconductors. Proc. 2-th Intern. Conf. Columbia, Missouri. April 23-26. 1978. Ed. J. M. Meese. N. Y. – London. Plenum Press. 1979. p. 249-259.

53. Senes A., Sifre G., Breant M. Stabilization of Transmutation Doping Silicon. In book: Semiconductor Silicon. 1977. Eds H. Huff, E. Sirtl. Princeton. The Electrochem Soc. p. 135-141.

54. Gres'kov I.M., Smirnov B.V., Solov'ev P.P., Stuk A.A., Kharchenko V.A. *Vliianie rostovykh defektov na elektrofizicheskie svoistva radiatsionno-legirovannogo kremniia* [The influence of growth

defects on the electrophysical properties of radiation-doped silicon]. Semiconductor Physics and Technology. 1978. Vol. 12. Iss. 10. p. 1879-1882. (in Russian)

55. Kharchenko V.A., Solov'ev P.P., Voronov I.II., Kuz'min I.I., Smirnov B.V. *Issledovanie metodom travleniia defektnoi struktury kremniia, obluchennogo bystrymi neitronami* [The etching study of the defective structure of silicon irradiated by fast neutrons]. Semiconductor Physics and Technology. 1971. Vol. 5. Iss. 4. p. 730-735. (in Russian)
56. Logan R.A., Peters A.J. Effect of Oxygen on Etch-Pit Formation in Silicon. J. Appl. Phys. 1957. Vol. 28. Issue 12. p. 1419-1423.
57. Pankratz J., Sprague J., Rudee M. Investigation of Neutron Irradiation Damage in Silicon by Transmission Electron Microscopy. J. Appl. Phys. 1968. Vol. 39. p. 101-106.
58. Voronov I.N., Gres'kov I.M., Grinshtein P.M., Guchetl' R.I., Morokhovets M.A., Sobolev N.A., Stuk A.A., Kharchenko V.A., Chelnokov V.E., Shek E.I. *Vliianie sredy otzhiga na svoistva radiatsionno-legirovannogo kremniia* [Effect of annealing medium on the properties of radiation-doped silicon]. Letters to ZhTF. 1984. Vol.10. Iss.11. p. 645- 649. (in Russian)
59. Svistel'nikova T.P., Moiseenkova T.V., Danilova N.O., Stuk A.A., Kharchenko V.A. *Vliianie na elektrofizicheskie svoistva kremniia primesei, prodiffundirovashikh pri iadernom legirovanii* [Effect on the electrophysical properties of silicon of impurities diffused during nuclear alloying]. Proceedings of the USSR Academy of Sciences. Inorganic materials. 1989. Vol. 25. Issue 1. (in Russian)
60. Moiseenkova T.V., Svistel'nikova T.P., Stuk A.A., Alontsev P.A., Kharchenko V.A. *Obratnaia diffuziia zolota i zheleza pri termoobrabotke v srede kislorod+khlor* [Reverse diffusion of gold and iron during heat treatment in oxygen + chlorine]. Proceedings of the USSR Academy of Sciences. Inorganic materials. 1990. Vol. 26. Issue 1. (in Russian)
61. Sobolev N.A., Stuk A.A., Kharchenko V.A., Shek E.I., Minenko P.V. *Vliianie sredy otzhiga na elektrofizicheskie parametry radiatsionno-legirovannogo kremniia* [Effect of annealing medium on the electrophysical parameters of radiation-doped silicon]. Proceedings of the USSR Academy of Sciences. Inorganic materials. 1990. Vol. 26.

Issue 8. (in Russian)
62.Nashel'skii A.Ia. *Tekhnologiia poluprovodnikovykh materialov* [Semiconductor Material Technology]. Moscow. Metallurgy. 1972. 432 p. (in Russian)
63.Guldberg J. Electron Traps in Silicon Doped by Neutron Transmutation. J. Phys. D: Appl. Phys. 1978. Vol. 11. p. 2043-2057.
64.Mukhin K.N. *Vvedenie v iadernuiu fiziku* [Introduction to Nuclear Physics]. Moscow. Atomizdat. 1965. 720 p. (in Russian)
65.Gordeev I.V., Kardashev D.A., Malyshev A.V. *Iaderno-fizicheskie konstanty* [Nuclear Physical Constants]. Moscow. GovAtomizdat. 1963. (in Russian)
66.Efimovich O.N., Solov'ev P.P., Stariznyi E.P., Stuk A.A., Sumin V.V., Kharchenko V.A. *Raspredelenie medlennykh neitronov v poli- i v monokristallicheskikh obraztsakh kremniia* [Distribution of slow neutrons in poly- and single-crystal silicon samples]. Atomic Energy. 1980. Vol.49. Iss.3. (in Russian)
67.Brugger R., Yelon W. In book: Pros. Conf. "Neutron Scattering", Catlinburg. 6-10 june 1976. Vol.11. p. 1117.
68.Gleston P., Edlung M. *Osnovy teorii iadernykh reaktorov* [Fundamentals of the theory of nuclear reactors]. Moscow. Publisher of Foreign Literature. 1954. (in Russian)
69.*Fizika iadernykh reaktorov* [Physics of nuclear reactors]. Translation from English edited by I.A. Stenbock. Moscow. Atomizdat. 1964. (in Russian)
70.Janus H., Malmros O. "IEEE Trans. Electronic Dev." Vol. ED-23, Issue 8. p. 797
71.*Normy radiatsionnoi bezopasnosti NRB-76* [Norms of radiation safety NRB-76]. Moscow. Atomizdat. 1978. 55 p. (in Russian)
72.Stone B., Hines D., Gunn S, McKown D. Detection and Identification of Potential Impurities Activated by Neutron Irradiation of Czocchralski Silicon. In book: Neutron Transmutation Doping in Semiductors. Proc. 2-th Intern. Conf. Columbia, Missouri. April 23-26. 1978. Ed. J. M. Meese. N. Y. - London. Plenum Press. 1979. p. 11-26.
73.Svistel'nikova T.P., Kharchenko V.A., Stuk A.A. Proceedings of the USSR Academy of Sciences. Inorganic materials. 1987. Vol. 23. Issue

1. (in Russian)
74. Mazel' E.Z., Afonin L.N. *Moshchnye kremnievye vysokovol'tnye tranzistory* [Powerful Silicon High Voltage Transistors]. Moscow. 1971. 75 p. (in Russian)
75. Dzhon G.F. *Problemy kachestva kremniia kak iskhodnogo materiala dlia sozdaniia silovykh priborov* [Quality problems of silicon as a source material for creating power devices]. Proceedings of the Institute of Electrical and Electronics Engineers. 1967. Vol. 55. Issue 8. p. 7-32. (in Russian)
76. Haas E., Schnoiler M. Phosphorus Doping of Silicon by Means of Neutron Irradiation. IEEE Trans. Electr. Dev. 1976. Vol. ED-23. Issue 8.
77. Pasynkov V.V., Chirkin L.K., Shinkov A.D. *Poluprovodnikovye pribory* [Semiconductors]. Moscow. Higher school. 1973. 398 p. (in Russian)
78. Platzoder K., Loch K. High-Voltage Thyristors and Diodes Made of Neutron-Irradiated Silicon. IEEE Trans. Electr. Dev. 1976. Vol. ED-23.
79. Kharchenko V.A., Solov'ev P.P., Novgorodtsev R.B. *Izmerenie potoka medlennykh neitronov s pomoshch'iu effekta Kholla v kremnii* [Slow neutron flux measurement using the Hall effect in silicon]. Atomic Energy. 1970. Vol. 28. Iss. 3. p. 253. (in Russian)
80. Tekhnicheskie usloviia. *Germanii i kremnii monokristallicheskie. Chetyrekhzondovyi metod izmereniia udel'nogo soprotivleniia* [Specifications. Germanium and silicon are single-crystal. Four-probe resistivity measurement method]. TU 48-05-4-162-71. (in Russian)
81. Batavin V.V., Kuznetsov Iu.N., Perezhogin G.A., Pantuev V.S, Popova G.V. *Primenenie radiatsionnogo legirovaniia dlia izgotovleniia odnorodnykh etalonnykh obraztsov kremniia* [The use of radiation doping for the manufacture of homogeneous reference samples of silicon]. Factory laboratory. 1972. Vol. 38. Issue 7. p. 819-820. (in Russian)
82. Gundorin A.P., Kuznetsov Iu.I., Shevel'kov M.A. *Standartnye obraztsy po udel'nomu soprotivleniiu i po tolshchine epitaksial'nogo sloia* [Standard samples for resistivity and epitaxial layer thickness]. Elektronnaia promyshlennost'. 1974. Iss. 9. Issue 33. p. 32-33. (in

Russian)
83.Ballantine D. Process Radiation Developments 1964-1965. Isotopes and Radiat. Techno! Vol. 2. Issue 2. p. 149-157.
84.Paaen H. Fabrication of Semiconductor Devices and Silicon Microcircuits by Neutron-Transmutation Dopintg. Isotopes and Radiat. Technol. 1970. Vol. 8. Issue 1. p. 37-60.
85.Klahr C, Cohen M. Neutron Transmutation Doped Semiconductors. Nucleonics. 1964. Vol. 22. p. 62-65.
86.Patent of USA N 3. 255. 050. 1966. Fabrication of Semiconductor Devices by Transmutation Doping. Clahk K.
87.Patent Issue 1442930 of 30.VIII.76 year. kl. VO1 17/00, UK.
88.Patent Issue 3967981 of 06.VII.76 year. kl. NO1 21/263, USA.
89.Patent Issue 2427645 of 02.01.76 year. kl. N01 17/34, USA.

Printed in Great Britain
by Amazon